DICTIONARY
OF
EVOLUTIONARY

·F·I·S·H·
OSTEOLOGY

ALFONSO L. ROJO

CRC Press
Taylor & Francis Group
Boca Raton London New York

CRC Press is an imprint of the
Taylor & Francis Group, an **informa** business

To my grandchildren

Brian, Patrick, Danielle, and Eric Alexander

ACKNOWLEDGEMENT

I am very pleased to express my gratitude to those ichthyologists, who have helped me with their ideas, criticisms and corrections. Since it is difficult to evaluate their individual contributions, I would like to name them in alphabetical order : Mr. Tom Amorosi (Hunter College. N.Y.); Dr. Laszlo Bartosiewicz (Archaeological Institute. Budapest); Dr. Norbert Benecke (Zentralinstitut für Alte Geschichte und Archäeologie. Berlin); Dr. Dirk Heinrich (Institut für. Haustierkunde. Kiel); Dr. Don McAllister (National Museums of Canada. Ottawa); Mr. Mark Rose (Senior Editor, Archaeology. New.York); Dr. W.B.Scott (Huntsman Marine Laboratory. Canada); Dr. Dale Serjeantson (London University. U.K.); and Dr. Wim Van Neer (Musée Royale de l'Afrique Centrale. Belgium).

The list of German terms has been checked by Dr. Angela von den Driesch (Institut für Palaeoanatomie. München), Dr. R. Nahrebecky (Saint Mary's University, Halifax, Canada), and by Drs. D. Heinrich, and N. Benecke. The Russian vocabulary has been checked and corrected by Dr. Yuri Glazov of Dalhousie University (Halifax, Canada), Dr. Yuri Riazantsev and Dr. A. N. Kotlyar both from VNIRO, Moscow, and Dr. Sergey Kovalev, from PINRO, Murmansk, USSR.

One person, Dr. Anthony Farrell (Saint Mary's University. Halifax. Canada), stands above all of them in my gratitude, for his constant and always friendly coaching in the correction of the English manuscript.

All drawings are original unless stated otherwise. Some of them (# 6, 7, 11, 14, 25, 26, 36, 42, and 43) have been used in a previous work (*Diccionario Enciclopédico de Anatomía de Peces*) published by the Instituto Español de Oceanografía, Madrid (Spain), from which I received its gracious permission to reproduced them here. I would like to give special thanks to my son Eric Rojo (E.R.) and to L. Niven for their assistance with the drawings and photographs, respectively. The cover and the line drawings signed A. R. are original work of the author.

I must emphasize that no one mentioned should be responsible for any errors. The last decision has been mine in all matters biological, artistic, and linguistic.

PREFACE

Like all language, zoological nomenclature [and we can add osteological nomenclature] reflects the history of those who have produced it. Some of our nomenclatural usage has been the result of ignorance, of vanity, obstinate insistence on following individual predilections of national customs, prides, and prejudices. Biological nomenclature has to be an exact tool that will convey a precise meaning for persons in all generations.

J.Chester Bradley
Preface to the 1st edition of the International Code of Zoological Nomenclature. 1961

Scientific knowledge is a continuum in constant revision. Facts, theories and interpretations are changing, with false conclusions shed rapidly from use. Although new generations have to adjust to such changes, they are managed without great difficulty. One branch of Biology, however, remains encumbered by the accumulation of dead weight. Nomenclature, the branch concerned with the assignment of names to new biological entities (whether anatomical elements, organisms or group of organisms), is fundamental to accurate scientific communication. Old names are difficult to discard, since they have entered and are well entrenched in the literature and vocabulary of science. The resistance to revision becomes more acute since it results from many factors, ranging from the lack of new scientific techniques to personal or national pride.

The scientific terminology applied to animals was taken care by the International Commission on Zoological Nomenclature when it adopted Linnaean binomial nomenclature. A set of rules revised periodically by the General Asembly of the International Union of Biological Sciences has put some order in Zoology. The International Anatomical Nomenclature Committee has proposed in different meetings a set of terms, under the name NOMINA ANATOMICA, to uniform medical terminology. For other biological branches, such as Osteology, Physiology, Behavior, etc. the consensus for a universally adopted vocabulary is still missing.

Within the nomenclature of fish osteology, there are at present hundreds of names given to the bones of fishes in different countries. The confusion of synonyms has been accentuated by some specialists' tendency to give new names based on the principle of homology. Although sound in principle, this approach is tainted by subjective interpretation. Binomial or trinomial terms have been proposed for a bone, known already by a simple, traditionally widespread name.

There is a need here to adopt a criterion that will create a uniform nomenclature, eliminating erroneous names and useless synonyms. Interpretation of bone features based on embryological, evolutionary, and physiological evidence are valid, but do not need to be incorporated in the nomenclature. This task should not be too difficult since the number of fish bones, although large, is limited.

This DICTIONARY was conceived as a result of many years of teaching and research on fish osteology, thanks to an awareness on the part of an increasing number of ichthyologists of the need for a comprehensive presentation of this topic, and on the part of archaeologists who have stressed the value of fish osteology to understanding the cultural importance of fish remains. This work is addressed to those interested in fish osteology from a professional point of view. Many students of fish osteology have been bewildered by the plethora of names given to fish bones, many of them, obviously, synonyms.

The DICTIONARY offers a rationale for the understanding of the nomenclature of all fish skeletal structures of modern fishes. Moreover, it provides an understanding of skeletal, morphological, and functional units, such as the suspensorium, the branchiae, and the caudal skeleton. Since the definition of osteological terms alone does not do justice to the complexity of ichthyology, the names of the most important taxa, together with those terms related to ichthyological methodology, have also been incorporated and explained.

The DICTIONARY describes the preparation of fish skeletons and gives the translation of each term in five languages (French, German, Latin, Russian, and Spanish). Drawings have been added to clarify the concepts defined in the DICTIONARY, to indicate the relative position of the bones in the fish skeleton, to show their anatomical and functional relationship, and to give some idea of the generalized shape of each bone and its variability within a common pattern. *The drawings are not intended to be used for identification purposes.*

Alfonso L. Rojo
Saint Mary's University
Halifax, N. S. Canada. B3H 3C3

CONTENTS

1. INTRODUCTION

A. Objectives

The purposes of the present DICTIONARY are to define the osteological and taxonomic terms referring to fishes, and to explain the rationale, both anatomical and functional, of fish skeletal units, in an evolutionary and biological context. This branch of biology --fish osteology-- provides a deeper insight into fish evolution, bone homologies, fish terminology, and fish taxonomy.

As a result, the main focus of this book is biological, and consequently it can serve as a guide for those interested in evolution, marine and freshwater biology, anatomy, paleontology, zoology, ichthyology, zooarchaeology, as well as teachers. Clearly, these fields overlap to a greater or lesser degree. Besides some are more theoretical, others more practical in nature.

Bones are the only anatomical parts that remain for a considerable time after the death of the fish. As such, they are witness of past existence and repositories of an enormous amount of biological and ecological information.

Retrieving this information is the task of the paleontologist and archaeologist, in collaboration with other specialists in disciplines as far removed from biology as meteorology and chemistry, among others. The results obtained can benefit many students in other fields related to biology.

One contribution of this book is to facilitate communication among specialists through emphasizing accurate definitions and eliminating doubts about the terminology of fish osteology. A common complaint of colleagues interested in fish osteology is the difficulty in interpreting the bone names given in older fish literature and in papers published in languages other than English. Both classes of fish literature often contain confusing bone synonymy The present DICTIONARY will greatly benefit those students entering their own special fields for the first time.

Biologists dealing with present marine and freshwater fishes are fortunate in having the required biological information directly available in most cases. Identification of the species, age, sex, size, weight, place, and environmental conditions are all easily obtained. Paleontologists and archaeologists, on the other hand, have to estimate data on live fish by indirect methods, drawing their information from fossilized or preserved bones.

In theory, a bone, any bone, can provide valuable and practical information for any of the specialists mentioned above. Depending on the bone and the information required, each specialist will profit to a different degree.

The reliability and abundance of the information obtained depends on the knowledge of the specialist, the methodology followed and the state of preservation of the material.

The following list presents the type of information that, directly or indirectly, fish bones can offer:

1. Identification of individual skeletal elements (bone, scale, spine, otolith) as fish components;
2. Recognition of their position in the fish body;

3. Recognition of the morphological features of the skeletal structures that bear information on age, sex, and geographical origin;

4. Identification of the fish family, genus, and species;

5. Calculation of the biological age of the fish;

6. Identification of the sex and sexual maturity of individual fishes;

7. Determination of the time of year of its capture or death;

8. Estimation of the total length or standard length of the live fish, by means of regression coefficients previously otained or by other less accuate method;

9. Estimation of the total or dressed weight of the live fish, directly from regression coefficients between bone dimensions and fish weight or indirectly fron length-weight keys, or simply by comparison with a reference collection;

10. Identification of the population in geographical terms, whether it is autocthonous, migratory or brought to the place by trade;

11. Reconstruction of the environmental conditions of the area or body of water (climate, aquatic habitat of the fish, interspecific and intraspecific relationship of the fish, etc.);.

12. Calculation of the MNI (minimum number of individuals) based on the comparison of the characteristics of the bones recovered (size, side in the fish, age, sex, etc.) in the area where the bones are found;

13. Estimation of the dietary value of the fish meat in calories, grams or pounds, or any other appropriate units;

14. The description and explanation of the postmortem manipulation of the fish skeleton (cutting, charring, burning, chewing, gnawing, etc.);

15. The study of the taphonomic circumstances that affected their distribution and modification (water and animal displacement, weathering, etc.); and

16. Recogniton of the nature of the bone deposits, either natural or human-related;

Obviously, not every bone provides all the information listed. Moreover, the estimation obtained depend on the state of preservation of the particular bone studied.

B. Criteria used in the selection of terms described in this DICTIONARY

The names applied to fish bones run into the hundreds. Many are synonyms, some are erroneous, some are misnomers, some are only known and used in a particular country, and finally, some have been found to be consistent with the principle of homology and therefore have been retained without change up to the present.

When selecting the osteological terms, it was immediately seen that other non-osteological names could and should be included because of their intimate relationship to the fish skeleton. In this category, terms related to biological methodologies have been included.

With these ideas in mind, the most useful names related to the fields listed below, have been selected from the enormous repertory of biological nomenclature, for the students of fish osteology

1. Taxonomy. Names of the main taxa of fishes, for example: elasmobranchs and ostariophysans;

2.   Nomenclature. Only monomial terms have been defined, such as *quadrate* and *suspensorium*, while most binomial or trinomial names based on embryological or evolutionary criteria are eliminated. Many of the latter, such *intertemporo-supratemporal-squamosal* are so cumbersome that for practical purposes should be eliminated.

3.   Synonymy. Synonyms used at present or in recent times have been included, while old names already discarded in modern works were excluded, for example: the term *ala minor ossis sphenoidei*, applied to the pterosphenoid;

4.   Osteology. Names of individual cartilages and bones, for example: Meckel's cartilage and quadrate. Also included here are the teeth and the otoliths, which although not part of the skeleton, are nevertheless very useful as a source of biological information.

5.   Morphology. Names of the different morphological parts of a bone, for example: condyle and process;

6.   Anatomy. Names of organs directly related to some skeletal elements, for example: ear capsule and lateral line;

7.   Comparative anatomy. Names of the types of scales, otoliths, vertebrae, and teeth, for example: ganoid scales, asteriscus otolith, opisthocelic vertebra, and heterodont teeth;

8.   Physiology. Names of the functional skeletal units, for example: suspensorium and branchiae.

9.   History. Names of bones most widely accepted by modern ichthyologists. The bones marked in the DICTIONARY by an asterisk, are those recommended by the author as official names; and

10.   Methodoly. The live features of fish, such as, length, weight, age, and sex, are useful in archaeological faunal studies and therefore terms referring to them are defined, as well as the names of techniques associated with them, for example, age reading and weight calculation.

C. Plan of the DICTIONARY

The remaining part of the GUIDE is divided into several sections. First is the SYNOPTIC SECTION, which contains a short index list of the main divisions into which fall all terms defined in the Descriptive Section. The second, the ANALOGICAL SECTION, offers all the terms arranged according to each of the categories listed in the synoptic section. This second section provides a double advantage to the reader. On the one hand, it offers an opportunity to increase the reader's vocabulary and to refine concepts or themes, by simply reading through the list of terms corresponding to a specific topic. But its most useful advantage is to provide an opportunity to the reader to readily find a word temporarily forgotten. In this case, a fast perusal of the listed words will suffice to recognize the word desired. If any doubt still persists between two terms, a consultation of the descriptive section should clarify the situation. The divisions and their sequence used in this section are those commonly accepted in courses on fish biology. Some words appear two or more times under different headings, whenever that concept corresponds to several different divisions. For example, the term *teeth* appears in the divisions MORPHOLOGY, INTEGUMENTARY SYSTEM, and DIGESTIVE

SYSTEM. Finally, the third and largest division, the DESCRIPTIVE SECTION, consists of the words defined at length.

Osteological and taxonomic terms are presented in alphabetical order, since this arrangement facilitates their reading and reference.

All the terms are given in English, in spite of the custom of using Latin terms in many English textbooks for some bones. The names of bones are given in singular when there is one bone only, either in the middle of the body or on its side (illicium, opercle). If there are two or more, either along the mid-line or on each side of the body, the name is given in plural (hypurals, branchiostegals).

The length of the descriptions depends on the importance and complexity of the anatomical structure concerned. Here, I have encountered a difficulty, i. e., how to present ideas related to different units without repeating *ad nauseam* the same concepts. The following example will clarify this point through the following series of integrated terms related anatomically and logically:

Skeleton
Axial skeleton
Skull
Neurocranium
Otic region
Prootic

The above terms are defined in such a way that the concepts referred by them are presented in an increasingly refined and specific way. The most general concept *skeleton* , includes a table indicating the units forming this system. The following term *axial skeleton* is defined and its units and elements are mentioned, and so on, until the last term, *prootic*, whose definition and specific characteristics are given. Whenever possible, any definition shows to which larger unit of the skeleton the term belongs and the smaller units or elements of which it is comprised. Here, the individual bones and cartilages are considered elements, while their higher groupings are designed as units (suspensorium, gill arches).

In this way, it is possible either to restrict the consultation to one term or to gather all information relevant to a topic. The definitions are, so to speak, interrelated, and provide an opportunity to pursue a topic from word to word. This approach combines the concepts of dictionary and textbook. With this in mind, it could be easily understood that it was sometimes impossible to avoid repeating certain explanations.

To supplement the definition, the following elements were added when appropriate: the classification of the structure according to different criteria; its etymology; the name of the anatomist who proposed the term; synonymy; references to illustrations and tables, and finally bibliography pertinent to the term.

Every term has been translated, when possible, into French, German, Latin, Russian, and Spanish. To my knowledge, only two extensive lists of fish bones have been published: one in French (Courtemanche and Légendre, 1985) and the other in Spanish (Rojo, 1988). Obviously, in any of these languages, there are several synonyms for every organ or concept, although generally only one translation has been provided. All translations were taken from original works and all have been checked by biologists from their respective countries. Singular and plural forms are given for some languages, when these forms differ widely. The singular or plural of Latin terms is not repeated in the lines corresponding to other languages, even when they use the same term.

Although many words derive from Greek, only Latin or latinized forms have been used here. According to modern trends, the Latin diphthongs æ and œ, have been simplified to e, for example: hemal for hæmal or opisthocelous for opisthocœlous. According to universal usage, however, the names of the fish families, similar to other animal and plant families, retain the diphthong, with both vowels typed separately. e. g., Gadidae and Cyprinidae. Latin words (apophysis, cranium) used in scientific English form the plural according to the rules of Latin grammar (apophyses, crania).

The next section is the BIBLIOGRAPHY where references are given in full. The bibliographical references are either included within the definition entries or added at their end in abbreviated form (author and year). Many well known biological principles have not been substantiated by bibliographical references, in order to avoid unnecessarily increasing the size of this section. The references provided can be used as a source of further readings.

The following section consists of the GLOSSARIES. The first glossary is made up of the English terms and includes the words not defined in the DICTIONARY, but referred to in the definition of other more important terms, where they can be easily found. This glossary is followed by the French, German, Latin, Russian, and Spanish lists, where the terms are given in their corresponding alphabetical order, preceded by the number corresponding to their English equivalent in the DESCRIPTIVE SECTION.

In German scientific terminology, the letter K is sometimes interchangeable with C, for example Kondylus or Condylus. Similarly, the letter Z is also many times replaced by C, for example, zycloid or cycloid. Also, the diphthongs "ae" and "oe" are often writen "ä" and "ö", respectively. Moreover many terms have two forms, one academic, usually taken from Latin (Apophyse) and the other vernacular (Auswuchs). When found, both have been given; if neither was available, the corresponding space for the name was left blank.

A similar situation occurs in scientific Russian, for example, обонятельная кост and этмóид for ethmoid. Since graphic accent does not exist in the Russian language, it is impossible to know which syllable should be stressed in a word. Moreover, many compound words, have usually two stressed syllables. In this DICTIONARY, the stressed vowel has been indicated with a graphic accent.

The final section is made up of the ILLUSTRATIONS (photographs and drawings). The drawings of individual bones and osteological units are oriented in their natural position in the fish. It is customary in biological works to consider right and left the sides of the fish when viwed in a swimming position. This interpretation has been followed in this book. The drawings are intended here only as illustrations of the large variety of shapes within a common pattern. They are not intended for identification purposes. Unless otherwise stated, all drawings are original. All drawings based on those of other authors, have been redrawn. For the bones represented in the illustrations, the maximum length or the maximum dimension (for bones of very irregular shape), is given in brackets. Although there is some difference in the shape of bones from fishes of different age and sex, in general, the shape remains constant during the adult stages.

2. FISH OSTEOLOGY

A. The fish skeleton plan

The supporting tissues of Vertebrates, known collectively as the skeleton, cover a wide gamut of anatomical structures, from the softest connective tissue to the hardest material in the body, tooth enamel. The function of the skeleton is the protection and support of the soft organs (heart, brain, and viscera), and to serve as attachments to the striated muscles. The supporting tissues are as follows: connective tissue and its types, notochord, cartilage in its different forms, the variable bony structures, and enamel.

Depending on the ontogenetic level of an individual fish, the skeleton is made up of either connective tissue, notochord, cartilage, bone, or some combination of the above. In early stages of development (embryos and larvae) the connective tissues, notochord and cartilage are most frequently found, while in the adult stage, the last two predominate.

The fish skeleton, the oldest of the Chordate phylum, is a complex mixture of elements that emerged in evolution as a result of the interaction of the fish with their environment. The structure of the fish skeleton has been the blueprint for the skeletons of other vertebrates, including that of man.

The fish skeletal elements of greatest interest are restricted to bones, teeth, spines and scales. Otoliths, although not skeletal structures, should also be included, since their resistance to decay, makes them useful as a primary source of biological information.

One of the major problems in the study of the fish skeleton is due to the plasticity of the group PISCES, which includes the most primitive vertebrates. Some forms such as the Cyclostomata, are also included in the group PISCES. The debate over whether lampreys are fishes or not, is left as a moot question. The fish group is very labile and has produced more forms than all the remaining vertebrate classes combined. Fishes alone, according to a conservative estimate, account for some 20,000 to 25,000 species, although some taxonomists raise this number to 40,000 species. All the remaining vertebrates combined barely exceed 20,000 species. A breakdown by classes accounts for approximately, 2,500 amphibians; 6,000 reptiles; 8,600 birds, and 4,500 mammals, of which man is one.

The study of the skeleton is difficult due to its progressive development, not only from embryo to adult (ontogeny), but also from the most primitive forms to more advanced ones (phylogeny). There is a continuum in both kinds of development, with a variable number of elements whose nature, shape, function, and number are changing through time.   The whole process is canalized into two opposing tendencies: specialization and degeneration that lead to dead evolutionary ends.

The study of the fish skeleton is necessary for a complete understanding of the main functions of the fish: locomotion, feeding, breathing, predation, and even reproduction in some families (Phallostethidae and Poeciliidae, among others). It also gives clues as to the species, sex, age, and geographical origin of the specimens. Moreover, several morphological and meristic characters of the skeleton have been used to identify and classify some taxonomic groups, whose names reflect this approach (Actinopterygii, Ostariophysi).   The evolutionary trend in the appearance and development of some of these features provides a clue for the understanding of the position of the fish in the evolutionary scale. Fishes with a cartilaginous skeleton

(Chondrichthyes) are primitive in relation to those with an ossified skeleton (Osteichthyes). More advanced bony fishes (acanthopterygians) usually have fewer vertebrae than lower bony fishes (malacopterygians).

But the most incontrovertible reason for the study of the skeleton in this context of our DICTIONARY is that in palaeontological and archaeological studies, the skeleton is the only remain of the fish we have at our disposal.

One good method to arrive at an understanding of the fish skeleton is to dissect one and reconstruct it again, as one does with a mechanical gadget. In the section (PREPARATION OF FISH SKELETONS) is given a procedure for this purpose.

The fish skeleton can be divided into two main large units according to its evolutionary history: the *endoskeleton* and the *exoskeleton,* also called *dermoskeleton* . As their names imply, the endoskeleton was located deep in the body, while the exoskeleton, external and of dermal origin, derives from superficial bones, covered solely by the epidermis in primitive fishes.

Both endo- and exoskeleton are loose terms without a clear distinction in practice. Some bones in modern fishes are the result of the fusion of two elements, one endochondral, belonging to the endoskeleton, and another dermal or exoskeletal. The palatine, for example, is the result of the fusion of an endochondral bone, the *autopalatine* and a dermal one, the *dermopalatine.*

The present DICTIONARY provides an understanding of the fish skeleton, by reading the entry *Skeleton,* followed by those entries referring to the names of the main skeletal units and their bones, as presented in table 5 in the same entry.

B. Fish osteological nomenclature

The naming of bones of the actino-pterygian head skeleton (and pectoral girdle) has resulted in a rather rich synonymy, most of which has not been directed toward understanding their homology. The central idea of comparative anatomy is the concept of homology...but most ichthyologists have been content to accept a system of names that serves descriptive and systematic purposes
Malcolm Jollie
A Primer of Bone (1986)

Fish osteological nomenclature has not attained yet the level reached by the terminology used in other sciences, such as physics and chemistry. Even Linnean binomial nomenclature is better organized since it follows clearly established rules.

Fish osteological nomenclature is in a state of flux, especially when compared to that applied to higher vertebrates. Although this is not the place for a history of fish nomenclature, it should suffice to offer two brief explanations, one historical and the other biological, to help understand the present situation.

From an *historical* point of view, fishes were one of the last vertebrate groups to attract the attention of anatomists. Although Aristotle and Pliny dealt at length with fishes, they showed no interest in the fish skeleton. Anatomical terms were first applied to the human body and later to those animals more useful to man (domestic and wild mammals) without knowing the homology among their skeletal parts. It is obvious that placental mammals and birds are very homogeneous groups, as far as the skeleton is concerned. There is variety in the shape and size of their bones, but not so much in their number and structure, while fishes, the vertebrate group most distant from mammals, have a skeleton with an infinite array of arrangements in the 20,000 to 25,000 species now living, not including fossil forms, which surpass by far this number. Moreover, the number of bones in the fish skeleton outnumber those of the reptilian, avian, or mammalian skeletons.

From a *biological* point of view, the study of the fish skeleton is difficult to interpret since many bones are the result of the fusion of several during the course of their embryonic development, as many as eight in the case of vertebrae, and three, at least, for the dentary, to give only two examples. This embryological reality is reflected in the larvae of advanced fishes and even in the adult stages of primitive fish groups. *Amia,* for example, has complete vertebrae (holospondylous condition) in the anterior section of the vertebral column, but in the posterior section every vertebra has two centra or, to be more accurate, two hemicentra (diplospondylous condition).

For an embryologist, a bone is defined by the number of ossifications in the cartilaginous or fibrous template of the embryo. An anatomist will define a bone according to its function and its relative position in the body. An evolutionist will consider a bone as the result of all the stages following from the first evolutionary stage in fossil fishes until its expression in an adult modern fish. All three approaches (embryological, anatomical, and evolutionary) respond to distinct dynamic criteria whose concepts cannot be crystallized in words, since they are static by nature. This linguistic problem is general to all branches of biology. The human zygote, from a pure biological point of view, is as human as an adult individual of the human species. Nevertheless in legal and ordinary life situations it is rather strange to call a zygote a *human*.

I have omitted the varied interpretation of diverse anatomical schools regarding the evolutionary or embryological origin of each skeletal element, retaining only the most widely accepted interpretation. The systematist and nomenclaturist and with them students of the fish skeleton have no other alternative than to contemporize and accept one view or another based on practical considerations.

To this nomenclatural problem based on diverse points of view, we have to add that ichthyology, as any other discipline, is an active intellectual endeavour which advances through error and new insights. New techniques help solve old problems, but they in turn create new ones. One of them is the readjustment of old terms to the new interpretations.

The standardization of fish bone nomenclature is a task that requires solution. Unfortunately, the number and variety of species, extinct and extant, together with the variability of the fish skeleton, makes this job extremely difficult, not to say impossible.

Three main nomenclatural difficulties immediately arise from this situation. The first one, is the result of the large number of osteological terms. Some terms, already obsolete, are found only in the classical works of the anatomists of previous centuries, while new ones are continuously coined by specialists when common anatomical terms already in use, could not be applied to the specialized structures found in the vast array of modern fishes.

The numerous elements in the fish skeleton compounded by the tendency to the multiplication of ossification centers in some fishes (*Lepisosteus*) or to reduction by fusion of several bones in others (higher teleosts), make the comparison of skeletons between different species and higher vertebrates very difficult and sometimes misleading.

The concept of homology introduced by Owen (1848) partially solved the problem. This approach allowed the possibility of giving one name only to the "same" bone in different groups. The advent of evolutionary theory by Darwin, expanded to fishes by T. H. Huxley and W. K. Parker, both rationalized the issue with the study of embryonic forms, and complicated it because of the lack of appropriate techniques and fossil data.

At present, there is an ongoing revision of the developmental stages of each bone, in an attempt to homologize them with those of primitive fishes and modern vertebrates. The result is less than satisfactory. On one hand, many actinopterygian bones (angular, dentary) have a double origin or are the result of the fusion of several bones. Those specialists interested in bone development look for names that reflect the true evolutionary development of the bone, and consequently, have proposed cumbersome names, such as *angulo-articular, dento-splenial*, not to mention *splenial-dentosplenial* or the monster term *dento-splenial-mentomandibular*.

On the other hand, some names already accepted by the majority of anatomists were proven wrong when their homologies were established. The term *alisphenoid* was rejected in favour of the new one *pterosphenoid*. One exception to this rule can be made in favour of three names, since they are already well established, although erroneously, in the fish literature. The names of these three bones are: a) the *vomer* , which should be called prevomer, since it is not homologous to the mammalian vomer, which has nomenclatural priority; b) the fish *frontal* is really the parietal, and c) the *parietal,* which should be called postparietal. These three names, or at least the last two, can be retained, based on historical prescription.

We should not be apprehensive in accepting new terms when the old ones have been proven to be incorrect on the basis of a study of bone homologies. No one will object today to the names, *supracleithrum, cleithrum*, and *postcleithrum*, although these bones were called epiclavicle, clavicle, and postclavicle (Cope, 1890) before the lack of homology between the cleithrum and the tetrapod clavicle had been established.

A second difficulty, a corollary of the first, stems from the immense variety of synonyms which are applied to the same structure. The "same" bone was given different names in different fishes, because of its variation in shape and position. This difficulty originates in the fact that past anatomists, working in isolation, did not know that a specific bone had already been named, and consequently they proposed a new one. This problem of synonymy was accentuated in some instances by national pride. With rare exceptions, I have selected the terms most widely accepted by present leading ichthyologists.

The third, a linguistic problem, has emerged from the historical development of anatomy. The first scientific works were written in Latin. Romance languages (Spanish, French, and Italian) changed the Latin terms according to their specific linguistic rules. These names, in the new form, are now used both in common and scientific language. Anglo-Saxon, Nordic, and Slavic languages, on the other hand, usually did not translate these names, and even now they are used in scientific papers in their Latin form. In Russian works, the text, written in Cyrillic characters, is often interspersed with the names of bones written in Latin characters.

As a basis for a possible uniformity of the osteological nomenclature I would like to propose two principles:

1) to accept a mononomial nomenclature avoiding compound names and descriptive phrases; and

2) to use the Latin terms of the bones or their derivatives in modern languages (e.g. operculum, opercle, opercule, opérculo). This approach seems more appropriate for an international endeavour than the use of vernacular names. In the case of Romance languages in which the translated names are so close to the original Latin that there is no mistaking their identity, the use of the vernacular offers no special problems. Nevertheless, in this DICTIONARY I have arranged the bones according to their English names.

3. PREPARATION OF FISH SKELETONS

Several methods are used in ichthyology to study fish skeletons. The advantage for the student rests on their didactic value for a first-hand understanding of the fish skeleton. The two most common and accessible to everybody are staining and softening the specimen in hot water. A third method, the cleaning of the skeleton by dermestid insects requires special care and facilities.

The staining process is a welcome complement to a reference collection of fish bones, collection which can be better prepared with the second method described later. This method produces best results with whole small specimens (larvae and juveniles) when the skeleton is already calcified. It also can be applied to parts of larger fishes (gills, skull, caudal skeleton).

The objective is to stain the calcified skeleton using alizarine red S, which gives it a beautiful red or purple colour. Cartilage can be stained blue using Alcian blue, methylene blue, or toluidine blue. A combined method (Dingerkus and Ohler, 1977) gives very good results in young specimens.

This method requires several steps which can be divided into four stages: fixing, digesting, staining, and clearing the specimen, but before starting, it is convenient to prepare the "stock" and "working" solutions according to the following or similar DICTIONARYlines.

A. "Stock" solutions

#1.   "Stock" KOH solution

KOH   0.5 to 1 part
Distilled water   99.5 to 99 parts
#2.   "Stock" sodium borate solution

Sodium borate in distilled water until saturation

#3.   Alizarin Red S

Alizarin in distilled water until saturation

#4. "Stock" dye solution

Alizarin solution	25 parts
Glacial acetic acid	5 parts
Glycerine	10 parts
Chloral hydrate (1 % solution)	60 parts

B. "Working" solutions

#5. "Bleach" solution

3 % Hydrogen peroxide	10 parts
"Stock" solution (#1)	90 parts

#6. "Buffer" solution

"Stock" sodium borate solution (#2)	30 parts
Distilled water	70 parts

#7. "Stain" solution

"Stock" dye solution (#4)	1 part
"Stock" KOH solution (#1)	99 parts

#8. 4 % KOH solution

KOH	4 parts
Distilled water	96 parts

#9. Glycerine I

Glycerine	20 parts
"Working" KOH solution (#8)	3 parts
Distilled water	77 parts

#10. Glycerine II

Glycerine —	50 parts
"Working" KOH solution (#8)	3 parts
Distilled water	47 parts

#11. Glycerine III

Glycerine	75 parts
Distilled water	25 parts

#12. Glycerine IV

Glycerine	pure

A.   Staining methods

1. <u>Staining bony tissue</u>

1.1. Fixing specimens

Step 1. Kill the specimen in 10 % formalin or 70 - 95 % ethyl alcohol. Leave it in solution for 7 - 10 days. Then add 1/2 teaspoon of sodium borate (borax) powder per quart of solution. When one wishes to be humane or to follow Government animal research regulations, it is recommended to kill the fish after being anesthetized by placing them in a mild solution of MSS 22, BioCalm or similar products.

Step 2. Soak all formaldehyde or alcohol from the specimens in running water, usually for 24 hours. Store the specimen in 70 % alcohol if it is not to be cleared immediately. For large specimens remove the viscera and the eyeballs, because they interfere with a clear view of some skeletal areas. Eyeballs should be preserved and stained, since in some cases the sclerotic layer of the eye ossifies into two or several bones.

Step 3. Put the specimen in "bleach" solution (#5). Leave no longer than 48 hours for big fish (more than 4 in) or 24 hours for smaller, except in rare cases when the amount of pigment requires longer periods.

1.2. Digesting the muscle tissue

Step 4. Place the specimen directly in sodium borate "buffer" solution (#6), usually for 24 hours.

Step 5. Add about 1/4 teaspoon of enzyme powder (pancreatine, pancreatine protease, or purified trypsin). Add more protein for larger animals.

Step 6. Change the solution (steps 4 and 5) in about one week or 10 days. Repeat until only a few areas of opaque muscle remain. This period, which varies for each fish, can be from one to several weeks.

1.3. Staining the specimen

Step 7. Place the specimen in "stain" solution (#7). Leave until all fin rays and bones can be seen clearly.

1.4. Clearing the specimen

Step 8. Place the specimen in 4 % KOH solution (#8) and leave for 24 hours or after the material no longer discolors the KOH solution. Repeat as long as necessary.

Step 9. Place the specimen in the glycerine series (I, II, III, and IV) for about 24 hours each (see "working" solutions #9, #10, #11, and #12).

Step 10. Keep permanently in glycerine IV. Add a few crystals of thymol to prevent growth of mould on the specimen.

2. Combined method for the staining of cartilage and bone

This method, similar to that used for staining bone, requires one more stage (Staining cartilage). Follow this procedure:

2.1 Fixing the specimen

Same as before

2.2 Staining cartilaginous tissue

The specific steps required to stain cartilaginous tissue are:
1. immersion of the specimen for 2 to 3 hours in a mixture of 10 mg of Alcian blue 8GN or similar product; 80 ml of 95 % ethyl alcohol; and 20 ml of glacial acetic acid, until desired color;
2. transfer of the specimen to 95 % alcohol for 3 days;
3. transfer of the specimen through an alcohol series (75 %; 40 %; and 15 %) for 2 to 3 hours in each or until the specimen sinks in the fluid; and
4. transfer to distilled water for 2 to 3 hours or until the specimen sinks.

2.3 Digesting the muscle tissue

Same as before, changing the solution until bone and cartilage are clearly visible.

2.4 Staining the bones

Same as before.

2.5 Clearing the specimen

Same as before.

Bibliography

Brubaker, J. M. and R. A. Angus. 1984. A procedure for staining fish with alizarine without causing exfoliation of scales. *Copeia*. 1984 (4) : 989-990

Dingerkus, G. and L. D. Ohler. 1977. Enzyme clearing of alcian blue stained whole small vertebrates for demonstration of cartilage. *Stain Technology*. 52 (4) : 229-232

Taylor, W. R. 1967. Outline of a method of clearing tissues with pancreatic enzymes and staining bones small vertebrates. *Turtox news*. 45 (12) : 308-309

B.   Softening in warm water methods

1. Without chemicals

A reference collection should include, if possible, skeletons of both sexes, from various geographical areas, and for different age or size groups, especially if there is a difference in the shape or size of the bones.

The following method provides one of the easiest ways to start such a collection. The instructions given are applicable to fishes of a manageable size. Once the technique is mastered, fishes of different sizes can be prepared for different types of displays (disarticulated or articulated skeletons or parts thereof, otolith and scale collections, etc.).

For a faster and better result in the beginning, select a fresh specimen of medium size (25 to 40 cm). All characteristics of the fish (sex, geographical origin, season, etc.) which might either affect the skeleton, or which are of interest for further reference (total and standard length, total weight, coloration, number of rays of the dorsal and anal fins, etc.) should be taken before cleaning the specimen. In species that do not show external sexual dimorphism, the sex is recorded after opening the abdominal cavity and establishing the presence of testes or ovaries. The weight after evisceration (dressed weight) is also an important information used to estimate the nutritional value of the specimen. Scales should be taken, at least, from the center of the body above the lateral line or from the area recommended by the specialists of the fish group. Removal of the skin and flesh for fish of smaller size than the one mentioned above is not required.

The easiest method to prepare the skeleton is to place the fish for 10 to 30 minutes, depending on size, in warm water hot enough for the tissues to be removed easily, but not so hot as to make the job uncomfortable. Do not boil, since the bones will fall apart, making their finding and identification very difficult. Also, laminar bones will curl at the borders. Large fish can be cut into pieces, but avoiding damaging any bone. The head, though, should be kept in one piece together with the pectoral and pelvic fins, when the latter are close to the pectoral fins.

After the bones have been cleaned of blood and tissues, they still retain in some cases a certain amount of fat which should be removed. Several fatty-dissolving chemicals can be used to degrease bones. Among the most common are carbon tetrachloride, benzine, and even pure gasoline. Leave the specimens submerged in the fluid in the sunshine until all grease has been dissolved. Clear or replace the liquid when it becomes laden with grease or dirt. This operation should be done in a well-ventilated area or in a gas chamber, since the vapors are either inflammable or toxic.

Bones that have been completely cleaned of all blood and tissues do not need to be bleached. But if need be, a simple procedure is to put them in the sun until the right color is attained. A soft solution (3 to 10 %) of hydrogen peroxide can be used. This process should be watched carefully to prevent damage to the bones, due to long exposure.

2.   With chemicals

A similar procedure to that described in B, but adding household detergents (mild soap flakes, ajax, etc.) to the water in a concentration appropriate to the size of the specimen.

C. Procedure for the preparation of fish skeletons

Once the fish has been softened in warm water, place it on its right side (its head to your left). With tweezers, remove the bones by twisting them, one by one, check their position, clean them of any soft tissue or blood in lukewarm water, and place them on a piece of paper in the same position as they occupy in the fish. Only the bones from the left side of the head and the left paired fins need to be prepared carefully. The bones from the right side can be collected later in any order.

A methodical procedure is advisable for a clear understanding of the main functional units of the skeleton, as well as each individual bone relationship and position. Follow this or slightly different order in the preparation of the skeleton. Some curators write a catalog number on each bone. This can help avoid mix-ups when bones are used for comparative identification.

1. Circumorbital series (Fig. 1A). The circumorbitals are thin bones located immediately below the skin. Because of their delicate structure they should be removed with tweezers. Starting at the anterior part of the orbit (9:00 o'clock position) remove the lachrymal (#1), which is generally the largest of them all. Continue with the remaining bones (#2-6) around the orbit in a counterclockwise manner. Check for the presence of supraorbitals and sclerotics. These last bones are formed in the sclerotic of the eye.

2. Nasal bone. It is a superficial bone close to the nostril (Fig. A1, #7).

3. Mandibular arch (Fig. 1B). Starting at the anterior part of the upper mandible, lift the premaxilla (#1), maxilla (#2), and if present, the supramaxilla (one or two). Follow with the lower mandible, lifting the dentary (#3) and angular (#4). Separate them carefully. Meckel's cartilage is found on the inner face of the angular. The retroarticular (# 5) usually remains attached to the angular in the preparations.

4. Opercular series (Fig. 1C). Start with the upper bone of the opercular membrane, the opercle (#1), and then continue down with the subopercle (#2) and the interopercle (#3). The last bone to be lifted from this area, although it does not belong to this series, is the preopercle (#4). The branchiostegals rays, which belong to this series, are best removed with the hyal arch. (See 11).

5. Tabular series (Fig.1C, #5). This series is formed by a group of paired bones located superficially above the opercular membrane. They are variable in number, usually very small and thin, and consequently difficult to locate in small specimens. There are four tabular bones on each side of the codfish head

6. Suspensorium (Fig. 1D). After the removal of the mandibular arch and the opercular series, the suspensorium is already exposed. Starting at the anterior part of the palate remove the palatine (#1), the endopterygoid (#2), and the ectopterygoid (#3). Continue upward lifting the quadrate (#4), the symplectic (#5), the metapterygoid (#6), and the hyomandibular (#7).

7. Pectoral girdle and its fins (Fig.1E). Remove the posttemporal (#1) from the posterior part of the skull and then move down to the supracleithrum (#2), cleithrum (#3), and postcleithrum (# 4). Some fishes have more than one cleithrum. Mesial to them are the scapula (#5), coracoid (#6), and radials (#7). These last bones are best kept together with the fins rays. Some fishes have a mesocoracoid in this area.

8. <u>Pelvic girdle and its fins</u> (Fig. 1E, #7). Due to their proximity, the pelvic bones from both sides can be taken together. Only the fin rays from the left side need to be saved.

At this point, with both hands, carefully remove in one piece the hyal arch and the branchial arches for further dissection.

9. Separate the remaining part of the skull (braincase) from the first vertebra (atlas). If the water is not too warm, the braincase should come in one piece. It is advisable not to separate the bones, in order to preserve their relationship. Another skull can be prepared with the bones separated, in order to recognize their individual shape. (Fig. 2).

Clean the outside of the skull. Eliminate the brain tissue by shaking the skull in water or by driving a stream of water through its openings. Collect the largest otolith from each side (sagitta, asteriscus, or lapillus). They can be used later to estimate the age. There are three pairs of otoliths, but usually two pairs (right and left lapilli and right and left asterisks) are very small and difficult to find with this method. It is better to retrieve them from another skull, carefully breaking open the otic capsule.

10. <u>Hyal arch and adjacent bones</u> (Fig. 1F). Although the branchiostegals (#1) do not belong to the hyoid arch, it is best to remove them now, beginning with the dorsalmost (the largest) and ending with the smallest in front. Notice their relationship to the hyal arch (either attached to its outer or inner face). *Number them starting with the dorsalmost.* Lift individually the bones of the left hyal series (hypohyals (#2), ceratohyal (#3), and epihyal (#4) or, according to modern nomenclature, dorso- and ventrohyal, anterohyal and posterohyal, respectively. Dorsal to this last bone is the small cylindrical interhyal (#5).

Between the right and left branch of the hyal arch, there is a middle series of very small bones or cartilages, (Fig. 3) made by the basihyal (in front), followed by one to three basibranchials. The urohyal, which does not belong to the hyoid arch, is located at the base of the tongue.

11. <u>Branchial arches</u> (Fig. 4). This complex unit is made up of five arches, each one consisting in most cases of four bones: pharyngobranchial (in a dorsal position); epibranchial, ceratobranchial (the longest), and hypobranchial (the ventralmost). The last arch has one bone only, the ceratobranchial. As mentioned earlier, between the right and left branchial series, there is a middle series of very small bones and cartilages made by the basihyal (in front), followed by several basibranchials. It is more convenient to clean each arch of any soft tissue (integument, muscles, and vessels) with fine tweezers, maintaining all bones, right and left, together. Once the position and relationship of each bone is understood, another specimen can be used to prepare each component independently (Fig. C)

12. <u>Vertebral column</u>. Starting with the vertebra closest to the head, clean every vertebra. It is recommended to number all vertebrae, starting with the atlas, at least in one specimen per species.

13. <u>Caudal skeleton</u>. The caudal skeleton is formed by the last modified vertebrae and the bones supporting the caudal fin rays. Since it requires more careful study, the last 10 or so vertebrae of the specimen should be kept together, if possible, with the caudal skeleton. Very carefully, clean the flesh with a scalpel, keeping all bones *in situ.* Let it dry for a day or so, and then glue it to a piece of card stock. A disarticulated caudal skeleton, especially if large, can also be prepared and stored.

14. <u>Median</u> <u>fins</u>. One of the medial fins, the caudal, has been already cleaned when dealing with the vertebral column. The dorsal and anal fins should be cleaned now. In fishes with numerous fin rays, only the largest can be cleaned and preserved. The total number can be registered unless it is already known from biological works. These fin rays are supported by means of the pterygiophores, made up of one to three pieces. There is no need to prepare all pterygiophores. The first two or three anteriormost are the largest in each fin and should be preserved.

Bibliography

Anderson, R. M. 1932. Methods of collecting and preserving vertebrate animals. *National Museum of Canada. Bull.* 69 : 1-141

Colley, Sarah M. and D. H. R. Spennemann. 1987. Some Methods of Preparing Fish Skeletons in the Tropics. *J. of Field Archaeology.* 14 : 117-120

Egerton, C. P. 1968. Methods for the preparation and preservation of articulated skeletons. *Turtox News.* 46 (5): 156-157

Konnerth, A. 1965. Preparation of ligamentary articulated fish skeletons. *Curator.* 8 (4) : 325 - 332. New York.

Mayden, R. L. and E. O. Wiley (1984). A method of preparing disarticulated skeleton of small fishes. *Copeia.* 1984 (1) : 230 - 232

D. Use of living organisms

1. A simple method is to leave the specimen in dechlorinated water to decompose, through the action of the bacteria already in the body. The cleaning will be complete in a few days. The container should be closed to provide the bacteria with an anaerobic environment. A temperature of 45 º C is the most appropriate for quick results.

2. Ants have also been used to clean specimens, which should be in a container with openings large enough to allow the ants to pass, but small enough to prevent them from taking small bones out .

3. In marine coastal waters, sand fleas and aquatic invertebrates can clean small specimens submerged in midwater, within a few days. The specimen should be placed in such a way as to prevent the loss of bones.

4. The use of dermestid beetles (*Dermestes*), popularly known as "bacon beetles", to clean skeletons is a method much favored in museum work, because it produces very satisfactory results. However, there are some drawbacks, since it requires a confined area to prevent the insects from escaping. Dermestids will feed on other valuable materials like dried meat, drying hides, zoological specimens (especially those having wool, fur, or feathers). For this reason, it is not a recommended method for the amateur collector working at home .

Bibliography

Colley, Sarah M. and D. H. R. Spennemann. 1987. Some Methods of Preparing Fish Skeletons in the Tropics. *J. of Field Archaeology.* 14 : 117-120

4. SYNOPTIC SECTION

The skeletal elements listed in the Analogical Section and defined in the Descriptive Section of this DICTINARY can be grouped in the following general categories.

1.   Basic terms

2.   Morphology

3.   Integumentary system

4.   Appendicular system

5.   Skeletal system

6.   Digestive system

7.   Reproductive system

8.   Organ of equilibrium

5. ANALOGICAL SECTION

214 mesethmoid
282 preethmoid
     septomaxilla
294 proethmoid
323 septomaxilla

b) orbitosphenoid region

12 alisphenoid
     pleurosphenoid
43 basisphenoid
44 belophragm
211 meningost
243 orbistosphenoid
274 pleurosphenoid
297 pterosphenoid
     alisphenoid
     pterosphenoid

c) otic region

29 auditory capsule
     labyrinth
     otosac
31 autopterotic
32 autosphenotic
47 bony labyrinth
124 epiotic
     epioccipital
241 opisthotic
295 prootic
298 pterotic
299 pterotic spine
328 sphenotic
332 squamosal

d) occipital region

41 basioccipital
123 epioccipital
127 exoccipital
     tectum synopticum
237 occipital region
267 pharyngeal process
341 supraoccipital

5.7 The dermocranium and its
     bones

102 dermocranium

a) dorsicranial region

9 adnasal
33 azygost
106 dermosupraoccipital
     dermal supraoccipital
     parietooccipital
140 frontal
180 intercalar
225 myodome
228 nasal
     adnasal
     naso-postrostral
     premaxillo-nasal
     prenasal
230 naso-premaxilla
236 occipital crest
257 parietal
     aparietal skull
     lateroparietal skull
     medioparietal skull
     parietooccipital skull
258 parietooccipital
283 prefrontal
286 prenasal
     rostral
311 rostral
339 supraethmoid
     dermal ethmoid
     dermal mesethmoid
      rostral
     dermethmoid
     mesethmoid
350 tabular bones
     cervicals
     extrascapulars
     nuchals
     postparietals
     scale bones
     supratemporals

b) basicranial region

39 basicranium
290 prevomer
373 vomer

c) orbital region

15 anamestic bones
17 antorbital
     superciliary
73 circumorbitals
105 dermosphenotic
178 infraorbitals

6. DESCRIPTIVE SECTION

1. Abdominal fishes

    Fr: poissons abdominaux
    Ger:
    Lat: pisces abdominales
    Rus: брюхопёрые рыбы
    Sp: peces abdominales

    Taxonomic group created by Linnaeus in his *Systema naturae* (1758), in which he included those bony fishes having the pelvic fins in an abdominal position. This feature represents, in the evolution of fishes, the most primitive stage of the position of the pelvic fins. The abdominal fishes include, among others, salmon, trout, and herring.

    Table 3

2. Acanthopterygians

    Fr: Acanthoptérygiens
    Ger: Acanthopterygier; Stachelflosser
    Lat: Acanthopterygii
    Rus: колючепёрые рыбы
    Sp: Acantopterigios

    Those actinopterygians having, as the name implies, spiny fin rays. This division represents a more advanced stage in evolution than that of malacopterygians. The acanthopterygians, comprising 35% of all living fishes, diversified into some 7000 species (Cohen, 1970). The name Acanthopterygii was first proposed by Willughby (1686).

    Table 2

3. Acanthotrichs *

    Fr: acanthotriches
    Ger: Acanthotrichen (sing: Acantotrich)
    Lat: acanthotrichia (sing: acanthotrichium)
    Rus: акантотрихии (sing: акантотрихия)
    Sp: acantotricos

    Scientific name given to the spiny rays of the dorsal and anal fins.

    See   Rays

4. Acrodont teeth

    Fr: dents acrodontes
    Ger: acrodonte (= akrodonte) Zähne
    Lat: dentes acrodontes (sing: dens acrodontis)
    Rus: акродонтные зубы (sing: акродонтный зуб)
    Sp: dientes acrodontos

Teeth fixed onto the top surface of some dermal bones and attached to them by connective collagenous tissue reinforced with calcium salts. The maxillary and mandibular teeth are also kept in place by a bony piece situated between the tooth and the bone. Most fish teeth are of acrodont type.

5. Actinopterygians

Fr: Actinoptérygiens
Ger: Actinopterygier; Strahlflosser
Lat: Actinopterygii; Actinopteri
Rus: лучепёрые рыбы
Sp: Actinopterigios

Subclass of bony fishes comprising all fish with membranous fins supported by soft or spiny rays. This group was divided in earlier classifications into three superorders of very different size: chondrosteans, holosteans, and teleosteans, their names referring to the gradual ossification of the skeleton. Although these three divisions are no longer considered taxonomic units, the names are very often used in teaching and in fish literature.

The most common osteological features of the actinopterygians are:
1)   the presence of premaxilla and maxilla;
2)   palatoquadrate not fused to the skull;
3)   dorsal fin, either single or divided; and
4)   radials at the base of the pectoral fins not expanded, except in Polypteriformes.

Syn: Actinopts
Tables 2, 3, and 4

6. Actinosts

Fr: actinostes
Ger:
Lat: ossa actinosta (sing: os actinostum)
Rus: птеригофоры (sing: птеригофор)
Sp: actinostos

See   Radials

7. Actinotrichs   *

Fr: actinotriches
Ger: Actinotrichen (sing: Actinotrich)
Lat: actinotrichia (sing: actinotrichium)
Rus: актинотрихии (sing: актинотрихия)
Sp: actinotricos

Goodrich (1904) named the original unsegmented dermal rays of the fin skeleton, *actinotrichs* . They develop in teleostean embryos, sometimes remaining close to the border of the fin during the adult stage of the fish, but generally they are replaced by the *lepidotrichia*. Actinotrichs are also found in the adipose fin of adult Salmonidae, Siluridae, etc.

Garrault (1936) considers them to be homologous with the ceratotrichs of the cartilaginous fishes, on account of their keratinous nature. In fish studies, actinotrichs and ceratotrichs are often used synonymously.

8. Adipose fin

Fr: nageoire adipose
Ger: Fettflosse
Lat: pinna adiposa (pl: pinnæ adiposæ)
Rus: жировой плавник (pl: жировые плавники)
Sp: aleta adiposa

A small fleshy fin, made up of adipose tissue reinforced by actinotrichs, and located on the fish back. The adipose fin is found on the postero-dorsal part of the fish back, between the dorsal and anal fins of several primitive and generalized members of the teleost families, Salmonidae, Myctophidae, Siluridae, Osmeridae, and Argentinidae, among others.

The adipose fin is lost in most paracanthopterygians and in all acanthopterygians, except in some tilefishes, in which there is found on the nape a tab of skin and flesh, called the *predorsal ridge* . This ridge is considered by some anatomists as an adipose fin.

Fig: 6

9. Adnasal

Fr: adnasal
Ger: Adnasale
Lat: os adnasale (pl: ossa adnasalia)
Rus: адназальная (= околоносовая) кость
Sp: adnasal

The middle bone of the three present in the nasal region of *Lepisosteus* .

See   Nasal

10. Agnathans

Fr: Agnathes
Ger: kieferlose Fische
Lat: Agnatha
Rus: Бесчелюстные рыбы
Sp·   Agnatos

As their name implies, the agnathans lack mandibles and consequently all the anatomical elements associated with them (masticatory muscles, bones and teeth). Feeding is reduced during the larval stages to filtration of planktonic organisms (microphagy) and, in exceptional cases, during the adult stage of life of some species. In most cases, the adults feed by sucking the blood of other fishes, as in parasitic lampreys. Adult hagfishes (Myxinidae) are scavengers, feeding mostly on the soft and decomposed tissues of dying or dead fishes. Their bodies are anguilliform, although their anatomy is markedly different from that of eels, which are bony fishes.

Agnathans, being descendants of the oldest group of fishes (ostracoderms) possess primitive characteristics, as well as specialized and degenerate features of their own.

Among the primitive or plesiomorphic skeletal features, we can mention:
a)   the absence of mandibles and related elements;
b)   the lack of paired fins (pectoral and pelvic) and their corresponding skeletons;
c)   the lack of an occipital region in the cranial skeleton;
d)   the lack of true branchial arches, the skeleton of which is replaced by an elaborate cartilaginous network, called *branchial basket*, which is external to the branchiae and intimately joined to the cranial skeleton;
e)   the presence throughout the adult stage, of a uniform, cylindrical notochord, pointed in the caudal region; and
f)   the lack of true vertebrae, having only neural and hemal arches in the caudal region.

Some specialized characteristics are:
a)   a mouth adapted for a sucking type of feeding and
b)   conical horny teeth with adaptations for tearing body tissues;
The degenerate features related to the skeleton are:
a)   an integument lacking exoskeleton (scales), and
b)   a membraneous and cartilaginous endoskeleton.
It is not likely to find remains of agnathans in archaeological sites, except possibly for the horny teeth in recent deposits.

Syn: jawless fishes
Tables   2, 3, and 4
Bibl: Grassé (1958); Hardisty and Potter (1971)

11. Alar scales   *

Fr: écailles carinales
Ger: Flügelschuppen; Alarschuppen
Lat:
Rus: крыловидные чешуи
Sp: escamas alares

Elongated scales found at the base of the caudal fin of some Clupeidae (*Alosa, Sardina,* etc.). Their size is related to the characteristic speed of the fish in swimming. Fishes which move slowly lack alar scales.

12. Alisphenoid

Fr: alisphénoïde
Ger: Alisphenoid
Lat: os alisphenoidale
Rus: крылоклиновидная кость; алисфеноид
Sp: alisfenoides

Name given to the pterosphenoid in early anatomical works on fishes. It is recommended that the use of this name be discontinued, since it has been proven that this bone is not homologous with the alisphenoid of the mammalian skull, which has name priority.

Syn: pleurosphenoid (de Beer, 1937)
See   Pterosphenoid

13. Amphic(o)elous vertebra *

Fr: vertèbre amphicœlique
Ger: amphicoeler (= amphizoeler) Wirbel
Lat: vertebra amphicœlica (pl: vertebræ amphicœlicæ)
Rus: амфицельный (= двояковогнутый) позвонок
Sp: vértebra anficélica

Most bony fishes have vertebrae of the amphicelous type, i.e., vertebrae having a centrum with two concave surfaces, one anterior and the other posterior, giving each vertebra an hour-glass shape.

The vertebral centrum is made up of two hollow cones of bone joined at their apices and reinforced by laminae and bars of spongy bony tissue. A mass of continous notochordal tissue fills the two cavities of each vertebra, maintaining the continuity along the vertebral column through the foramen pierced in the center of each vertebral body. This type of vertebra is found in all elasmobranchs, *Amia,* and in all teleosts, except in gars (*Lepisosteus*).

Fig: 35 B

14. Anal fin

Fr: nageoire anale
Ger: Analflosse; Afterflosse
Lat: pinna analis (pl: pinæ anales); pinna ani (pl: pinnæ ani);
        proctopterygium (pl: proctopterygia)
Rus: анальный плавник (pl: анальные плавники)
Sp: aleta anal

The anal fin (*proctopterygium*) is located on the median plane of the body immediately behind the vent. It has a shape similar to that of the dorsal fin and like it functions to maintain equilibrium. Sometimes it is split into two or more units, as in Gadidae. The anal fin of acanthopterygians has two sections: one anterior, with spiny rays (*acanthotrichs*) and the other posterior, supported by soft rays (*lepidotrichs*).

Among the most striking modifications of the anal fin in some fishes is its transformation into a copulatory organ, which has been used for poeciliid classification.

Syn: proctopterygium
See: Gonopodium
Figs: 6 and 13

15. Anamestic bones

Fr: os anamestiques
Ger:
Lat:
Rus: анаместические кости; кости анамний
Sp: huesos anamésticos

Westoll (1936) defined anamestic bones as "those supernumerary bones present in the cheek region of fishes, that fill the free spaces left by the sensory pit-bearing bones." This led some authors to apply this term to all bones that lack sensory canals.

It seems that there exists a relationship between the lengthening of certain areas of the head (rostrum, cheek) and the number of new ossifications that cover the spaces between the preexisting bones.

Table 1

16. Angular *

Fr: angulaire
Ger: Angulare
Lat: os angulare (pl: ossa angularia)
Rus: угловáя кость
Sp: angular

Paired bone of mixed origin, partially endochondral, but predominantly membraneous, that forms the posterior section of the mandible. In teleosts the angular has a triangular shape with the anterior angle fitting between the two branches of the dentary. It articulates posteriorly with the quadrate.

Although many authors call it *articular,* Haines (1937) and Lekander (1949) showed that it should be called angular, since the membraneous part that constitutes the larger part of this bone corresponds to the true angular.

C. Reichert was the first to recognize the homology of the angular with the median ear ossicle, called hammer or *malleus* in mammals.

See   Retroarticular
Syn: dermarticular (Goodrich, 1930); articular (Gregory, 1933; Berg, 1940)
Figs: 1B, 3, 13, 18, and 25A

17. Antorbital *

Fr: antorbitaire
Ger: Antorbitale
Lat: os antorbitale (pl: ossa antorbitalia)
Rus: предглазнúчная (= анторбитáльная) кость
Sp: anteorbitario

Paired dermal bone located in front of the eye. It has sometimes been considered part of the infraorbital series, because of the infraorbital canal that crosses it.

The antorbital, located in front of the lacrymal, is found in   *Amia, Lepisosteus, Elops,* in Osmeridae, and in some Siluridae, among other osteichthyans.

Syn: adnasal; superciliary (Valenciennes)

18. Apode fishes

Fr: poissons apodes
Ger: fusslose Fische
Lat: pisces apodes
Rus: голобрюхие рыбы
Sp: peces ápodos

Group of bony fishes lacking pelvic fins. The absence of pelvic fins represents the most recent stage in the evolutionary migration of the pelvic fins towards the head, from a primitive position in the abdominal region, to their complete disappearance, as in the eel (*Anguilla*) and moray (*Muraena*).

Syn: apodal fishes
Table 3

19. Apophysis

Fr: apophyse
Ger: Auswuchs; Apophyse
Lat: apophysis (pl: apophyses)
Rus: апо́физ; вы́рост; отро́сток (pl:отро́стки)
Sp: apófisis

Any narrow expansion that protrudes from the body of a bone.

20. Appendicular skeleton

Fr: squelette appendiculaire
Ger: Extremitätenskelett
Lat: skeleton (= sceleton) appendicularis
Rus: коне́чностей скеле́т; скеле́т прида́тков
Sp: esqueleto apendicular

The appendicular skeleton comprises all the cartilages and bones that support the paired fins (pectoral and pelvic). In some descriptions, the skeletal structures of the median fins are also included as part of the appendicular skeleton.

In bony fishes, the appendicular skeleton has evolved into a complex system of many supporting elements for the fin rays. The pectoral skeleton is more highly developed than the pelvic. In some cases, due to a secondary evolutionary process, the pelvic skeleton of some fishes is reduced or disappears altogether. The most striking stage in the evolution of the appendicular skeleton is the loss of the pelvic fins and their skeleton in apode fishes (*Anguilla*).

Figs: 13 and 25 B

21. Articular

Fr: articulaire
Ger: Articulare
Lat: os articulare (pl: ossa articularia)
Rus: сочленовная (= артикуля́рная; = суставна́я) кость
Sp: articular

In primitive actinopterygians, an endochondral bone, is found, which occupies the same position as Bridge's ossicles *b* and *c* of *Amia.* The articular is later invaded by the angular. For practical purposes, this name is restricted to the few primitive North American fishes (*Amia, Lepisosteus, Polyodon,* and sturgeons).

Syn: anguloesplenial (Devillers, 1958); anguloarticular; angular
See   Angular
Bibl: Rojo (1986)

22. Articular process

Fr: processus articulaire
Ger: Articularfortsatz
Lat: processus articularis (pl: processus articulares)
Rus: сочленовный (= артикулярный) отросток [pl: с-ные (=а-ные) отростки)]
Sp: proceso articular

Process found on the upper border of the premaxillary, typical of the most advanced actinopterygians. It serves as a fulcrum for the head of the maxillary in its outward movement when the mouth opens.

Fig: 15

23. Articulation

Fr: articulation
Ger: Gelenk
Lat: articulatio (pl: articulationes)
Rus: сочленение (pl: сочленения); сустав
Sp: articulación

A point or plane of union between two bones. Articulations are classified according to two criteria: the amount of movement allowed between the two bones and the nature of the tissue that binds them.

The articulation that permits free bone movement is called a *diarthrosis*. *Amphiarthrosis* describes the type of articulation allowing little bone movement, e.g., the joint between two vertebrae in most fishes. The *synarthrosis*, permits little or no movement, as the joint between the two lower mandibles.

The bones are joined by connective or cartilaginous tissue. A joint effected by cartilage, such as the mandibular joint, is called *symphysis*, while a *suture* is a rugged line between two bones cemented by connective tissue. Sutures are common between the cranial bones in teleosts.

24. Ascending process

Fr: processus ascendent
Ger: aufsteigender Fortsatz
Lat: processus ascendens (pl: processus ascendentes)
Rus: восходящий отросток (pl: восходящие отростки)
Sp: proceso ascendente

This name is applied to the vertical expansion found in most teleosts on the anterior part of the premaxillary. The ascending process is also present in holosteans, [gars (*Lepisosteus*) and bowfin (*Amia*)]. According to Patterson (1973) these two processes are not homologous and consequently he proposes the term *nasal process* for that of the holostean bone.
Fig: 15

25. Aspondylous vertebra   *

Fr: vertèbre aspondyle
Ger: aspondyler Wirbel

Lat: vertebra aspondyla (pl: vertebræ aspondylæ)
Rus: аспондильна́льный позвоно́к (pl: а-ные позвонки́)
Sp: vértebra aspondila

Fish vertebrae result from the assemblage of several bony elements (neural spine, neural arch, centrum, hemal arch, hemal spine) which are added at different times during the embryogeny of the fish. Most primitive fishes (cyclostomes, dipnoans, *Acipenser,* etc.) have all these elements except the centrum, hence the name *aspondylous*, meaning without centrum, applied to their vertebrae.

26. Asterisk   *

Fr: astériscus
Ger: Asteriscus; Lagenolith
Lat: asteriscus (pl: asterisci)
Rus: слухово́й ка́мешек
Sp: asterisco

The asterisk is the otolith found in the lagena of actinopterygians. It is held in an upright position leaning against the crista of the lagenar macula, which is innervated by the auditory branch of the eighth cranial nerve, the stato-acoustic nerve. It is usually very small except in the Cyprinidae (carp family), in which it is the largest of the three otoliths. It is the last to appear during embryonic development.

Although of small size in many fishes, it has specific features, which can be of archaeological value.

Syn: lagenolith
Bibl: Shepherd (1910)

27. Asterospondylous vertebra   *

Fr: vertèbre astérospondyle
Ger: asterospondyler Wirbel
Lat: vertebra asterospondyla (pl: vertebræ asterospondylæ)
Rus: астероспондильна́льный позвоно́к (pl: а-ные позвонки́)
Sp: vértebra asterospondila

The type of vertebra found in cartilaginous fishes, characterized by the calcification extending into the chordacentrum and the autocentrum. In a transversal section, these calcified areas have a star-like appearance, while the areas corresponding to the arches remain cartilaginous, e.g., the vertebrae in the man-eater shark (*Carcharodon*).

Fig: 35 A

28. Atlas   *

Fr: atlas
Ger: Atlas
Lat: atlas
Rus: атла́нт; пе́рвый ше́йный позвоно́к
Sp: atlas

Name given to the first vertebra, which articulates directly with the skull. It has usually a strong neural spine that reinforces the connection of the vertebral column with the skull.

See   Vertebral column
Fig: 35 B

29. Auditory capsule

Fr: capsule auditive
Ger: Gehörkapsel
Lat: capsula auditiva (pl: capsulæ auditivæ)
Rus: слуховáя (= ушнáя) кáпсула [pl: слуховы́е (=ушны́е) кáпсулы]
Sp: cápsula auditiva

The internal ear of vertebrates, also known as *labyrinth,* is enclosed in the auditory capsule formed in the embryo from a pair of auditory placodes. Each placode soon develops into a large vesicle, the *otocyst,* with a narrow evagination on its dorsal region. On the nape of the head of elasmobranchs each evagination elongates and opens to the exterior by the *endolymphatic pore.* The evagination itself is called *endolymphatic duct.*

The otocyst later divides into two sections by a horizontal constriction, although both sections remain connected in most cases. The upper section forms the three semicircular canals and the utricle; the lower or vestibular section develops into the saccule and the lagena.

The auditory capsule is formed by cartilage in chondrichthyans, but in bony fishes, a series of chondral bones (prootic, opisthotic or its replacement, the intercalar, epiotic, also named epioccipital, sphenotic, pterosphenoid, and basisphenoid) form the walls and bottom of each capsule, while the parietals and frontals form its roof.

30. Autopalatine

Fr: autopalatin
Ger: Autopalatinum
Lat: os autopalatinum (pl: ossa autopalatina)
Rus: эндрохондрáльная нёбная (= автонёбная) кость
Sp: autopalatino

—   The autopalatine is an endochondral bone found in primitive bony fishes, which in later forms fuses with a dermal bone, with or without teeth. The resulting bone is named simply *palatine.*

See   Palatine

31. Autopterotic

Fr: autoptérotique
Ger: Autopteroticum
Lat: os autopteroticum (pl: ossa autopterotica)
Rus: автокрыловидноушнáя кость; автоптербтикум
Sp: autopterótico

Paired bone that constitutes the endochondral component of the pterotic.

See Pterotic

32. Autosphenotic

Fr: autosphénotique
Ger: Autosphenoticum
Lat: os autosphenoticum (pl: ossa autosphenotica)
Rus: автосфенотикум
Sp: autoesfenótico

Paired bone that forms the endochondral component of the sphenotic. In primitive fishes the autosphenotic was an independent bone.

See   Sphenotic

33. Axial skeleton

Fr: squelette axial
Ger: Achsenskelett
Lat: skeleton (= sceleton) axialis; (pl: skeletones [= sceletones] axiales)
Rus: осевой скелет (pl: осевые скелеты)
Sp: esqueleto axial

The axial skeleton, composed of the bones located in the middle line or axis of the fish body, can be subdivided into the four following units:
a)   the neurocranium, formed by the braincase;
b)   the branchial skeleton, formed by the hyoid arch, the branchial arches and the opercular membrane bones;
c)   the vertebral column with the vertebrae and the caudal skeleton; and
d)   the paraxial bones, i.e., intermuscular bones, and ribs.
In fishes, the primary dynamic role of the axial skeleton is its passive participation in locomotion, since it acts as leverage for the trunk musculature. This function changes to one of support in tetrapods.

Bibl: Stokley (1952)

34. Axonost

Fr: axonoste
Ger: proximales Segment der Flossenträger
Lat: os radiiferum (pl: ossa radiifera)
Rus: проксимальный элемент
Sp: axonosto

Cope (1890) gave this name to the proximal piece of the pterygiophore.

See   Pterygiophore
Syn: proximal pterygiophore
Figs: 13 and 41 C

35. Azygost

Fr: azygoste
Ger:
Lat:
Rus: непа́рная кость
Sp: azigosto

Name proposed by Chabanaud (1936) for the dermal bone in the family Psettodidae (Pleuronectiformes), located between the prefrontal and the frontal of the nadiral (lower) side.

36. Basapophysis

Fr: basapophyse
Ger: Basapophyse
Lat: basapophysis (pl: basapophyses); processus transversus (pl: processus transversi)
Rus: базапо́физ; базостро́сток
Sp: basapófisis

Lateral, small, paired bony expansions located on the lower part of the vertebral centra.

Syn   parapophysis
Fig: 35 C

37. Baseost

Fr: baséoste
Ger: distales Segment der Flossenträger
Lat:
Rus: диста́льный элеме́нт
Sp: baseosto

The distal segment of the pterygiophore, which act as support for the fin rays (Cope 1890).

Syn: distal pterygiophore; intercalarium (Brühl, 1877)
See   Epibaseost; Pterygiophore
Figs: 13 and 41 C

38. Basibranchials   *

Fr: basibranchiaux (sing: basibranchial)
Ger: Basibranchialia
Lat: ossa brasibranchialia (sing: os basibranchiale)
Rus: базибранхиа́льные (= основны́е жа́берные) ко́сти [sing: базибранхиа́льная (= основна́я жа́берная) кость]
Sp: basibranquiales

Nelson (1969) defines the basibranchials as the cartilaginous or bony median elements of the branchial skeleton, excluding the tooth dermal plates sometimes associated with the latter. Each basibranchial is numbered according to the branchial arch that follows it. The branchial arches of both sides articulate ventrally by means of the paired hypobranchials (I-IV), at the joining point of two basibranchials.

It is customary to call each one of the basibranchials, *copula*. Actinopterygian fishes have usually three copulae: the first one is called the *basihyal*, and the posterior almost invariably remains cartilaginous for life.

See   Basihyal
Figs: 4 and 25 A

39. Basicranium

Fr: basicrâne
Ger: Basicranium
Lat: basicranium (pl: basicrania)
Rus: базикра́ниум
Sp: basicráneo

Term used to designate, in a rather vague way, the bones of the ventral side of the neurocranium and the dermal bones associated with it.

40. Basihyal   *

Fr: basihyal
Ger: Basihyale
Lat: os basihyale (pl: ossa basihyalia)
Rus: языко́вая кость (pl: язико́вые ко́сти)
Sp: basihial

The basihyal is the anteriormost median endochondral bone of the basibranchial series. It joins both branches of the hyoid arch and forms the skeleton of the tongue in teleosts. Its dorsal surface is sometimes covered by a tooth plate of dermal origin, more appropriately called *glossohyal*. In some families (Salmonidae) the basihyal cartilage fails to ossify and remains cartilaginous for life.

It is customary to use the term basihyal to refer, in a more general sense, to the complex basihyal, dermal plate, and teeth.

Syn: basihyobranchial; dermal basihyal
Figs: 4 ans 25 A

41. Basioccipital   *

Fr: basioccipital
Ger: Basioccipitale
Lat: os basioccipitale (pl: ossa basioccipitali)
Rus: основна́я затыло́чная (= ни́жнезатыло́чная) кость
Sp: basioccipital

Median bone of endochondral origin that forms the posterior part of the base of the skull. It forms the ventral part of the rim of the *foramen magnum*. The basioccipital is located between the two exoccipitals and meets the parasphenoid anteriorly. It terminates with a concave facet, which articulates with the centrum of the first vertebra. In cyprinoid fishes, it expands posteriorly forming the *pharyngeal process*.

See   Lower pharyngeal
Figs: 2 and 36 A

42. Basipterygium   *

Fr: basiptérygie
Ger: Basipterygium
Lat: os basipterygium (pl: ossa basypterygia)
Rus: базиптерйгиум
Sp: basipterigio

The paired chondral bone that supports the pelvic fin in bony fishes. Both basipterygia join anteriorly at the pubic symphysis forming the pelvic girdle. In most cases, three expansions or processes are present: one anterior or *pubic* ; a middle one or *iliac*, and a posterior or *ischial process* . The body of the bone, known as the *pubic plate,* has a small depression, the *acetabular facet*, where the fin rays or the radial bones articulate.

See   Pelvic girdle
Syn: pelvic bone
Figs: 1 E, 13, and 41 A

43. Basisphenoid   *

Fr: basisphénoïde
Ger: Basisphenoid
Lat: os basisphenoideum (pl: ossa basisphenoidea)
Rus: основная клиновйдная кость; базисфенбидная кость; базисфенбид
Sp: basiesfenoides

Median bone of endochondral origin present in primitive bony fishes. It forms part of the floor of the neurocranium between the foramen of the internal carotid and that of the pseudobranchial artery.

In teleosts, it takes the shape of the letter "Y", the wings of which surround the pituitary gland. The basisphenoid forms the base of the posterior myodome. The basisphenoid arises from three centers of ossification: one medial, the *belophragm* of Chabanaud; and two lateral, the *meningosts,* that form the wings of this bone.

The basisphenoid remains in a cartilaginous stage in Ostariophysans and is lost in some teleosts (Gadiformes).

See   Belophragm and Meningost
Fig: 14 A

44. Belophragm

Fr: bélophragme
Ger:
Lat:
Rus:
Sp: belofragma

Name proposed by Chabanaud (1936) for the median ossification of the basisphenoid in salmon.

See   Basisphenoid

45. Beryciform foramen

Fr: foramen béryciforme
Ger: beryciformes Foramen
Lat: foramen beryciform
Rus:
Sp: foramen bericiforme

Name applied to the perforation of uncertain significance found in the ceratohyal of Beryciformes and some other groups. In some cases, the dorsal rim of the perforation is lost and only a notch remains on the dorsal edge of the ceratohyal bone.

Fig: 32

46. Bone

Fr: os (pl: os)
Ger: Knochen
Lat: os (pl: ossa)
Rus: кость (pl: кости)
Sp: hueso

Bones are the units which form the skeleton of bony fishes. Bony tissue, one of the hardest of the skeletal tissues in vertebrates, represents the final stage in the evolution of the supporting tissues, beginning with the mesenchyme of the embryo. Bone tissue is primarily a postembryonic tissue, in contrast to cartilage, an essentially embryonic structure.

The ground substance or *matrix*, typical of bone, results from the secretory activity of the bone cells, the *osteocytes*. The osteocytes are arranged in parallel layers maintaining communication among themselves by means of filamentous cytoplasmatic processess running perpendicularly to the layers. These cells disintegrate with time, but the cavities in which they were encased remain, preserving the shape of the cells. For this reason, this type of bone is called *cellular bone*. The matrix is hardened by the accumulation of calcium carbonate, but mostly by the abundance of calcium phosphate, which forms a compound called *hydroxyapatite*, the formula of which is [3 Ca_3 $(PO_4)_2$. Ca $(OH)_2$]. The cellular bone is typical of lungfishes, crossopterygians, chondrosteans, the most primitive teleosts (Salmonidae, Siluridae, Cyprinidae, and Anguillidae), and some genera of the most advanced teleosts (*Perca, Gadus, Thunnus,* etc.).

In *osteoid* tissue, described by Kölliker (1858), the osteocytes lack the characteristic ramifications of those of the cellular type. In the osteoid type, the osteocytes disintegrate and the lacunae fill with matrix, a feature that warrants the name *acellular type,* by which it is known. The acellular bone is present in the most advanced teleosts (*Mola, Cyclopterus,* etc.).

Bones can be also classified into different types according to their phylogenetic, ontogenetic, and osteogenetic origin; their position in relation to other bones; and their relation to teeth and sensory canals.

Any bone can be used to identify the fish and to estimate their live size. However, the error in the estimation will be larger when using small bones, since there will not be enough difference in their sizes and consequently the correlation coefficient will be small. The reason being that the measuring devices commonly used (rulers, calipers) do not accurately measure small linear units. Better result is obtained when measuring relatively large bones.

In order to get a statistically accurate estimation of the live size one needs regression formulae obtained from a statistically valid sample of fishes taken from the same population (Casselman, 1974; Fagade, 1974; Rojo, 1986; Scott, 1977; Van Neer, 1989). A common practice in archaeological studies is simply to compare the size of the subfossil bone with those of a modern collection.

The wide variety of Osteichthyes presents a vast array of bone shapes. The same bone, although built on a common pattern, presents such a variation of outlines in different species, that it is not easy to homologize their morphological features. Processes, crests, foramina, and spines are present in some fish groups while absent in others. As a result, the universal standardization of fish bone measurements is, at best, doubtful, if not impossible.

However, if the standardized technique of measurement is applied to related orders, or to clusters of related families within the same order, then standardization is possible, and obviously highly desirable and useful.

At present, as far as I know, there has been only one attempt (Morales and Roselund, 1979) at standardization of the measurements of the bones of four distant families (Gadidae, Cyprinidae, Esocidae, and Pleuronectidae). In that work, with the exception of the vertebrae, which offer no special problem, the dimensions selected from the remaining bones should be revised. In general, homology among specific bone features has yet to be established.

Table   1
Bibl: Cannon (1987); Casteel (1974a, 1976); Chaplin (1971); Jollie (1986); Mundell (1975); Olsen (1968); Yerkes (1977).

Table 1. Classification of bones according to diverse criteria

PHYLOGENY		ONTOGENY		OSTEOGENESIS		OTHER CRITERIA
1. dermal	=	1. membrane	=	1. achondral	=	1. covering bones
						2. dentigerous
						3. canal
						4. anamestic
2. non dermal	=	2. cartilage	=	2. chondral	=	5. replacement
				3. parachondral		
				4. perichondral		
				5. endochondral		
3. mixed (dermal and chondral)						6. sesamoid

See these terms

47. Bony labyrinth

Fr: labyrinthe osseux
Ger: knöchernes Labyrinth
Lat: labyrinthus osseus
Rus: ко́стный лабири́нт
Sp: laberinto óseo

The osseous enclosure within which the membraneous labyrinth is located. It is formed in front by the otic bones (prootics, pterotics, sphenotics, epiotics or epioccipitals, and opisthotics or their successors in evolution, the intercalars), at the rear by the occipital series (supraoccipital, exoccipitals and basioccipital), and on top by the dermal roof (frontals and parietals).

See   Membraneous labyrinth

48. Brachiopterygians

Fr: Brachioptérygiens
Ger: Brachiopterygier
Lat: Brachiopterygii
Rus: многопёрообразные рыбы
Sp: Braquiopterigios

Fish group of doubtful taxonomic position. The modern tendency is to place it within the actinopterygians, with which they share many anatomical features, except that their pectoral fins have a peculiar structure: they are lobate and their radials (propterygium and metapterygium) extend well into the fin. They are represented nowadays by the genus *Polypterus*, which includes ten species and the reedfish (*Calamoichthys malabaricus*).

49. Branchiae

Fr: branchies
Ger: Kiemen
Lat: branchiæ (sing: branchia)
Rus: жа́бры (sing: жа́бра)
Sp: branquias

See   Gills
Fig: 25 A

50. Branchicteniae *

Fr: branchicténies
Ger: Branchictenien; Reusenfortsätze
Lat: branchicteniæ (sing: branchictenia)
Rus:
Sp: branquispinas

Term proposed by Chabanaud to designate the gill rakers, because of their similarity to comb teeth.

See   Gill rakers

51. Branchiocranium

Fr: branchiocrâne
Ger: Branchiocranium
Lat: branchiocranium (pl: branchiocrania)
Rus: бранхиокра́ниум
Sp: branquiocráneo

The branchiocranium is the assemblage of skeletal units, cartilaginous or bony, related at one time or another, to the branchiae during fish evolution. In this context, its extension varies in different groups of fishes, being larger in cartilaginous fishes than in bony fishes.

52. Branchiopercle

Fr: branchioperculaire
Ger: Branchioperculum
Lat: os branchioperculum (pl: ossa branchiopercula)
Rus:
Sp: branquiopérculo; branquiopercular

Hubbs (1919) gave this name to the fourth bone of the opercular series of *Amia* , covered partially by the subopercle and the interopercle. Nelson (1968) considers this name to be superfluous, since this bone is simply the most dorsally located branchiostegal ray.

53. Branchiostegal rays   *

Fr: rayons branchiostèges
Ger: Branchiostegalstrahlen; Kiemenhautstrahlen
Lat: ossa branchiostegalia (sing: os branchiostegale); radii branchiostegi
Rus: бранхиостега́льные (=жа́берные) лучи; [sing: б-ный
      (=жа́берный) луч]; лучи жа́берной перепо́нки
Sp: radios branquióstegos

Bony fishes possess a typical branchiostegal apparatus formed by a series of long, curved and often pointed bones, called *branchiostegal rays* or simply *branchiostegals*, that support the branchiostegal membrane. The branchiostegals are dermal bones of variable shape and number, whose origin can be traced to the lateral gular bones of palaeoniscoid fishes. The wider end of the rays, its *head*, rests close on the ventral edge of the ceratohyal (*sensu lato*); the first ones are attached to its outer side, and the remaining ones to the inner face. Although in association with the hyoid arch, they really belong to the opercular series.

The shape of the branchiostegal rays varies widely: it is filiform in Anguilliformes and Catostomidae and spatiform or acinaciform in *Perca* . Their number varies in the different groups of modern bony fishes according to their phylogenetic origin, as follows:

I	Actinopterygians	0	-	50
	a) sturgeon (*Acipenser*)			1
	b) bowfin (*Amia*)	10	-	13
	c) gar (*Lepisosteus*)			3
	d) malacopterygians	0	-	36
	e) acanthopterygians	1	-	26
II	Crossopterygians			
	(*Latimeria*)			0
III	Dipnoans	0	-	3

The largest number of rays, found in actinopterygians and especially in their fossil forms, represents the first evolutionary stage. Note that, in spite of the fact that many modern actinopterygian species are the most advanced in evolutionary terms, this group, as a whole, is the oldest of the three mentioned above.

Because of the thinness and lack of distinctive marks, most branchiostegal rays are difficult to identify. Their number, a familial and generic trait, can be used to calculate MNI, especially for those species with a small number. This calculation is somewhat doubtful for those fishes having a large number of long, and consequently, fragile branchiostegals.

In some fish species (*Cristivomer namaycush*), branchiostegals have been used to determine the age of fish.

Figs: 1 F, 3, 6, 13, and 25 A
Bibl: Bulkley (1960); Gosline (1967); McAllister (1968)

54. Camptotrichs

Fr: rayons camptotriches
Ger: Kamptotrichen (sing: Kamptotrich); Fadenstrahlen
Lat:
Rus: камптотри́хии (sing: камптотри́хия)
Sp: camptotricos

Name coined by Goodrich (1904) to designate the segmented, branched, fibrous or slightly ossified fin rays of dipnoans and crossopterygians.

55. Canal bones

Fr: os à canaux ; os sensoriels
Ger: Schleimkanalknochen
Lat:
Rus: тру́бчатые ко́сти
Sp: huesos de canal

The canal bones, also called *sensory canal bones*, are those bones of dermal origin that enclose the neuromasts and the branches of the sensory canal system. The canal bones are formed from one ossification center, as in the case of the infraorbital bones, or from several centers, as in the nasal and frontal.

Syn: sense organ bones
Table 1

56. Caniniform teeth

Fr: dents caniniformes
Ger: caniniforme Zähne
Lat: dentes caniniformes (sing: dens caniniformis)
Rus: КЛЫКОВИ́ДНЫЕ ЗУ́БЫ (pl: к-ный зуб)
Sp: dientes caniniformes

The caniniform teeth are conical or elongated in shape and have a sharp end, either straight or curved. They are found in predator fishes, who use them to grasp, pierce or hold their prey. In many cases, the caniniform teeth are hinged, bending backwards to allow the entrance of the prey but snapping into a locked position when the prey tries to escape.

Figs: 12 A and B

57. Capsular ethmoid

Fr: ethmoïde capsulaire
Ger: Kapselethmoid
Lat:
Rus:
Sp: etmoides capsular

Paired bone of perichondral origin, located on the inner walls of the nasal capsule of some teleosts. It has a concave surface fitting against the walls of the capsules.

58. Cardiform teeth

Fr: dents en velours; dents en cardes
Ger: cardiforme Zähne; hertförmige Zähne
Lat: dentes setiformes (sing: dens setiformis)
Rus:
Sp: dientes cardiformes

Teeth characterized by being short, fine, pointed, and very numerous; they resemble the card used to prepare and clean wool. Many species belonging to the families Percidae (perches), Serranidae (sea basses), and Ictaluridae (freshwater catfishes) have cardiform teeth.

Fig: 12 B

59. Cartilage

Fr: cartilage
Ger: Knorpel
Lat: cartilago (pl: cartilagines)
Rus: ХРЯЩ (pl: ХРЯЩИ́)
Sp: cartílago

Each of the cartilaginous elements that, independently or associated, protects or supports body organs. The skeleton of cyclostomes and chondrichthyans is made up exclusively of independent cartilages (as for example the palatoquadrate and the annular cartilage), or by complex units, such as the neurocranium.

The statement which has appeared repeatedly in archaeological papers since Bryan (1963), who reported that according to Dr. A.D. Welander of Washington School of Fisheries, "salmon bones are cartilaginous and therefore not preserved", is erroneous. Bone and cartilage are two completely different tissues; no bone is cartilaginous, and cartilage cannot be considered a bone. These two tissues are fundamentally different at the cellular level. Nevertheless, some fish families (Salmonidae, Molidae) have spongy, soft or thin bones, which decompose easily.

60. Cartilage bones

Fr: os de cartilage
Ger: Ersatzknochen
Lat: ossa substituentia (sing: os substituentium); ossa chondrogena
Rus: замеща́ющие (=хондра́льные; =хрящевы́е) ко́сти
(sing: з-щая кость; х-ная кость; хрящева́я кость.)
Sp: huesos de cartílago

Bones formed by replacement of the cartilaginous tissue by bony tissue, through a process called *osteogenesis*. The cartilage bones can be called, either *parachondral, epichondral,* or *endochondral,* depending on whether the process starts in the connective tissue surrounding the cartilage, in the perichondrium, or directly inside the cartilage, respectively. Often one bone follows two of these paths in its osteogenesis, but the final result, including that of the achondral type, typical of the membranous bones, is always the same and it is impossible to find any structural difference in the adult bones. Only a detailed study of the embryology of the fish may cast light on the ontogenetic and phylogenetic origin of a particular bone.

In most cartilage bones (perichondral and parachondral types) this process can be divided into two stages: the *metaplasia stage,* characterized by the transformation of the connective tissue into cartilage, followed by the *neoplasia stage* , in which the cartilage is replaced by bone. In this last stage, osteogenesis is preceded by *chondrolysis* or destruction of the cartilage. The endochondral bones are formed by this last process exclusively.

Although it is true that there are difficulties in assigning a bone to one or another origin, in practice it is more expedient to reduce the classification of bones to two categories: membrane bones and cartilage bones. This last category includes the three mentioned above. In the case of bones of mixed origin, they can be assigned to either group depending on the predominant type.

Syn: replacement bones; chondral bones
Table 1

61. Caudal bony plate

Fr: plaque osseuse caudale
Ger: Deckknochen der Chorda
Lat:
Rus: хвостова́я ко́стная пласти́на
Sp: placa ósea caudal

Name given to the first pair of uroneurals, much larger than the remaining ones, that is located on the curve made by the upturned vertebral column. It is preferably to call it *first uroneural,* since it can be followed by one or two uroneurals. Salmonidae have three uroneurals which are never fused with the centra, a condition that can occur in other fishes. The presence of three uroneurals is a more primitive condition than having one or two.

This bone, by its peculiar shape, can be a good detector of salmonid fishes.

62. Caudal fin

Fr: nageoire caudale
Ger: Schwanzflosse
Lat: pinna caudalis (pl: pinnæ caudales); uropterygium (pl: uropterygia)
Rus: хвостовóй плавнйк
Sp: aleta caudal

The median fin located at the end of the body forms the tail (*uropterygium*) in most fishes, but is absent in fishes with elongated bodies. The caudal fin supplements the propulsive force exerted by the body (and other fins), increasing their hydrodynamic efficiency.

Syn: uropterygium
Figs: 6, 13, and 37

63. Caudal peduncle

Fr: pédoncule caudal
Ger: Schwanzstiel
Lat: radix caudæ (pl: radices caudæ); pedunculus caudalis (pl: pedunculi caudales)
Rus: хвостовóй стéбель
Sp: pedúnculo caudal

It is that part of the fish between the end of the base of the anal fin and the beginning of the caudal fin. Its height has been used as a morphological trait to characterize species or races. The number of scales below the lateral line in the caudal peduncle has been used as a specific feature in some biological works. Figure 6 shows the recommended method of counting these scales.

Fig: 6

64. Caudal skeleton

Fr: squelette caudal
Ger: Skelett der Schwanzflosse
Lat: skeleton caudalis
Rus: хвостовóи скелéт
Sp: esqueleto caudal

The caudal skeleton of fishes, sometimes called *urophore,* is a complex unit made up of bones derived from either cartilage or dermal tissues. Because of its continuous and vital activity in fish locomotion, the tail has been exposed to profound morphological and structural modifications through evolution. This circumstance makes the tail of fishes a very useful tool in fish classification and fish interrelationships.

The most primitive (though hypothetical tail) is known as *protocercal* or *proterocercal tail* having a straight axis and simmetrically arranged elements decreasing in size towards the end.

An evolutionary trend in tail development produced a tilting of the vertebral axis (notochord or vertebral column), either upwards, typical of modern elasmobranchs, or downwards present in some ostracoderms. Many modern fish embryos still show the upward tilt of the axis at the extremity of the body. This abbreviated heterocercal tail can be seen in embryos and larvae of many teleosts, and in the adults of chondrosteans (sturgeons, paddlefishes), even when the tail is externally homocercal, as in the holostean *Amia* .

The caudal skeleton of osteichthyans consists of a series of bony elements of either dermal or endochondral origin, arranged according to the following pattern. It should be noted that not all these elements are found in all fishes.

A. EPAXIAL ELEMENTS

1. *Dermal elements*

a)   *Fulcra* (sing: fulcrum): a median series of forked scales, variable in number, riding the anterior edge of the tail in primitive fishes.

b)   *Urodermals*: a double series of small bones variable in number, derived from scales, according to Nybelin (1963) and Patterson (1968).

c)   *Caudal rays*, which include the simple small caudal rays at the upper and lower edges of the tail, as well as the simple branched rays, either segmented or unsegmented.

2. *Endochondral elements*

a)   *Epurals*: a series of one to three median bones located dorsally to the urostyle and supporting the caudal fin rays.

b)   *Uroneurals*: paired structures of rodlike bones of variable number and representing neural arches.

B. AXIAL ELEMENTS

All axial structures are of endochondral origin, except for the tip of the notochord which persists as soft tissue in many forms.

a)   *Stegural*: a term proposed by Monod (1968) for the paired structure which articulates with the first preural centrum. According to Patterson (1968) it represents the first uroneural fused with the first preural and first ural vertebrae.

b)   *Preural centra:* the vertebral bodies preceding the bifurcation of the caudal artery.

c)   *Ural centra:* the vertebral centra found after the bifurcation of the caudal artery. Some of the last centra represent preural and ural vertebrae fused together in a variety of patterns.

C. HYPAXIAL ELEMENTS

1. *Dermal elements*

A similar series to that found on the epaxial region, but lacking the urodermals.

2. *Endochondral elements*

a)   *Hypurals:* a median series of fanlike bones below the urostyle and supporting the caudal rays.

b)   *Parhypural:* the hemal arch of the first preural centrum, which is the last hemal arch crossed by the caudal vein and artery. The parhypural sometimes has a process called *hypurapophysis.*

It should be noted that not all these elements are found in all fishes.

A recent study by Fujita (1989) deals with the terminology of the cartilaginous elements found in the caudal skeleton of teleostan fishes

Figs: 13 and 37
Bibl: Affleck (1950); Barrington (1935); Buhan (1972); Gosline (1961); Lundberg and Baskin (1969); Nybelin (1963, 1973).

65. Ceratobranchials   *

Fr: cératobranchiaux
Ger: Ceratobranchialia (= Keratobanchialia)
Lat: ossa ceratobranchialia (sing: os ceratobranchiale)
Rus: кератобранхиа́льные ко́сти (pl: к-ная кость)
Sp: ceratobranquiiales

Paired bones of endochondral origin belonging to the branchial arches and located between the epi- and hypobranchials. They are the longest bones of the gills. The last pair of ceratobranchials, the only branchial bones left in the fifth branchial arch, bears a tooth plate on its dorsal surface. Most fishes have five pairs of ceratobranchials, although they are missing in *Polypterus, Calamoichthys,* and in some Anguillidae.

The fifth pair of ceratobranchials in Cypriniformes and Siluriformes, known as the *inferior pharyngobranchial bone,* is the strongest bone of the branchiae and bears teeth arranged in rows. North American species have only one or two rows of pharyngeal teeth (Berg 1912), while in the rest of the world there are species with one, two, or three rows. Vladykov (1934) has studied the world distribution of cyprinid fishes in relation to the number of teeth rows on this bone.

See   Dental formula and Infrapharyngobranchial dental plate
Figs: 4, 25 A, and 33

66. Ceratohyal   *

Fr: 1. cartilage cératohyal
         2. os ceratohyal
Ger: 1. Ceratohyalknorpel (= Keratohyalknorpel)
         2. Ceratohyale (= Keratohyale)
Lat: 1. cartilago ceratohyalis (pl: cartilagines ceratohyales)
         2. os ceratohyale (pl: ossa ceratohyalia)
Rus: 1. кератохиа́льный хрящ (pl: кератохиа́льные хрящи́)
         2. кератохиа́льная кость (pl: кератохиа́льные ко́сти)
Sp: 1. cartílago ceratohial
         2. hueso ceratohyal

1. The ceratohyal cartilage is a paired element located on the ventral part of the hyoid arch of cartilaginous fishes.

2.   The homologous ceratohyal bone of bony fishes is therefore of endochondral origin, and as such has no sensory canals. The ceratohyal articulates

dorsally with the interhyal; its anterior border supports some branchiostegal rays and ventrally it joins one or two hypohyals.

Most fish anatomists (Holmgren and Stensiö, 1936) admit that in teleostean embryos, the ceratohyal together with the epihyal, correspond to two ossification centres of one same bone. According to this interpretation, the anterior, ventral or proximal ossification is given the name of *ventral ceratohyal* (known in the past *ceratohyal*) and the posterior, dorsal or distal ossification is named *dorsal ceratohyal*, called by previous authors (and still by many) *epihyal*.

This new nomenclature has created a new problem of homologies. If the ventral ceratohyal is homologous, as many suspect, to a hypobranchial, then it should not be called ventral ceratohyal. The same logic applies to the dorsal ceratohyal. To solve this problem, Nelson (1969) proposed the following names for the hyoid arch bones: *dorsohyal* and *ventrohyal* for the two hypohyals, because of their relative position; *anterohyal* for the old ceratohyal (or ventral ceratohyal); and *posterohyal* to replace the old epihyal.

For the sake of uniformity with most authors, the old names, ceratohyal and epihyal, are retained in this DICTIONARY

See   Hyoid arch
Figs: 1 F, 3, 25 A, and 32

67. Ceratotrichs

Fr: rayons cératotriches
Ger: Hornstrahlen; Keratotrichen (sing: Keratotrich)
Lat: ceratotrichia (sing: ceratotrichium)
Rus: кератотрихии (sing: кератотрихия)
Sp: ceratotricos

Name given by Goodrich (1904) to describe the non-segmented dermal rays of Elasmobranchii and their fossil relatives. Ceratotrichia are made of two parallel rods, sometimes ramified at the tip. Krukenberg established in the last century that the chemical composition of the ceratotrichia is a protein different from others known in fishes, to which he called *elastoidine*.

68. Chevrons

Fr: chevrons
Ger:
Lat:
Rus: гаммаобразные косточки
Sp: chevrones

The characteristic V-shaped scales located in the belly of some clupeids, giving a serrate profile to the belly of these fishes.

69. Chondrichthyans

Fr: Chondrichthyens
Ger: Knorpelfische
Lat: Chondrichthyes
Rus: хрящевые рыбы
Sp: Condrictios

Although the cartilaginous nature of most of the chondrichthyan skeleton does not generally lend itself to preservation, some skeletal elements (vertebrae, teeth, spines, and placoid scales) have nevertheless turned up in several southwestern archaeological sites having the right condition for preservation.

The chondrichthyans or cartilaginous fishes are characterized by the presence of a cartilaginous skeleton. Their most important osteological features are the presence of

a)   mandibles with teeth;

b)   vertebrae, sometimes with calcified areas; and

c)   placoid scales, covering the entire body in sharks, scattered or in rows in rays and skates, and scattered in the tenaculum, claspers, and midline of back in chimaeras.

Willughby (1686) established this division for the first time in his *Historia piscium,* with the name of *Pisces cartilaginei.* At present the chondrichthyans are divided into two subclasses: Elasmobranchii, comprising the sharks, rays, and skates, and the Holocephali, formed by the chimaeras.

Syn: cartilaginous fishes
Tabl.   2, 3, and 4.
Bibl: Applegate (1965); Gans and Parsons (1981)

70. Chondrocranium

Fr: Chondrocrâne
Ger: Knorpelschädel
Lat: chondrocranium (pl: chondrocrania)
Rus: хондрокра́ниум; хрящевóй чéреп (pl: хрящевы́е чéрепы)
Sp: condrocráneo

Chondrocranium is the assemblage of all cartilaginous elements of the head at any time in the life of the fish. During embryonic development, the chondrocranium comprises all the cranial cartilages that later will form the neurocranium and the viscerocranium. The dermocranial elements are never part of the chondrocranium, since they ossify directly from the dermal tissues.

The embryonic skull starts as a membraneous structure, which is soon changed into a cartilaginous one (chondrocranium), and it, in turn, is replaced by the osteocranium. As the fish grows, the chondrocranium decreases in extension until, in the adult of many teleosts, it is reduced to a small cartilaginous area between the bones, providing them with the means of growth and mobility.

In certain fishes (cyclostomes and chondrichthyans) the chondrocranium remains unchanged for life. It is called platybasic, when during the embryonic stages, the trabeculae are independent and the brain grows between them, as in elasmobranchs. It is tropibasic, or moderately so, when the trabeculae of the embryo fuse together forming a plate on which the brain rests, as in chondrosteans, holosteans, and teleosteans.

See   Platybasic and Tropibasic skull

71. Chondrosteans

Fr: Chondrostéens
Ger: Chondrosteer
Lat: Chondrostei
Rus: хрящекостúстые рыбы; хрящевы́е ганóиды
Sp: Condrósteos

Müller (1844) grouped together the sturgeons and the paddlefishes (*Polyodon*) under the name Chondrostei and separated them from the *Amia* and *Lepisosteus*, with which they had been grouped by Aggassiz (1833) under the common name of Ganoidei. The term chondrosteans refers to the nature, partially cartilaginous and partially bony, of their skeleton.

In modern classifications, the chondrosteans form the order Acipenseriformes which includes both sturgeons and paddlefishes. They are characterized by the presence of some primitive characters, such as rostrum, mouth in a ventral position and heterocercal tail. They have also, at the same time, some degenerate characters, such as the loss of the ganoid scales (present only in the caudal peduncle) and a cartilaginous skeleton with some ossifications.

Table 2

72. Ciliated scales

Fr: écailles ciliées
Ger: Cilienschuppen
Lat:
Rus: реснйчные чешу̃и
Sp: escamas ciliadas

Ctenoid scales, in which the spinules or ctenii are very long, as in the boarfish (*Caprus aper*) of the southern coast of Europe.

See   Ctenoid scales

73. Circumorbitals

Fr: circumorbitaires
Ger: Circumorbitalknochen (= Zircumorbitalknochen); Circumorbitalia
Lat: ossa circumorbitalia (sing: os circumorbitale)
Rus: бкологлазнйчные кости (sing: б-ная кость)
Sp: huesos circunorbitarios

The circumorbitals comprise an assemblage of small, thin, and transparent bones of dermal origin, forming the eye orbit, and surrounding more or less completely. the eye of bony fishes. The series of circumorbital bones is complete only in the most primitive bony fishes, *Lepisosteus* , and in some Osteoglossidae.

Although these bones have a varied phylogenetic origin and different characteristics, all are included in the term circumorbitals. Nevertheless, they receive different names according to their relative position in the orbit. The *antorbital* is located in the anterodorsal area of the orbit followed in a clockwise manner by the *supraorbitals, postorbitals, infraorbitals* and below them, the *suborbitals*. These terms are not absolute, as it is sometimes difficult to recognize the boundaries of the areas occupied by these bones. It is recommended that the name *suborbitals* be limited for the series located ventrally to the infraorbitals. The suborbitals are found in palaeoniscoid fishes and are not associated with the infraorbital sensory canal.

In certain groups of teleosteans (Myctophidae), the infraorbitals have a shelf that extends inward, acting as a support for the eye.

It is difficult to homologize each individual bone in the different groups of fishes, since their fragile consistency makes them an inappropriate material for fossilization.

Figs: 1 A and 14 B

74. Classification   of Fishes

Fr: classification des Poissons
Ger: Klassifikation der Fische
Lat: classium distributio Piscium
Rus: классификация рыб
Sp: clasificación de los Peces

Early classifications of fishes were established according to a morphological criterion, based generally on one or several of the most important characteristics, e.g., position of the pelvic fins, types of scales, presence or absence of a pneumatic duct, number of fins, etc.

These classifications were not successful, because even if the groups were identified, it was soon found that fishes with diverse characteristics, not to say incompatible ones, were included in the same group. The reason for this situation is that, in fish, as in any other animal group, evolution operates in a mosaic pattern, that is to say, each organ follows its own evolution independently of the remaining organs of the body. The eye, for example, has maintained an almost identical structure from fish to man, while the gills have undergone profound modifications until they have finally disappeared. There is, nevertheless, a certain degree of interdependence among related organs.

Another reason for the failure of these early classifications is the fact that morphological and anatomical similarity does not necessarily imply a phylogenetic relationship between their possessors. For example, the hydrodynamic shape of such unrelated animals as shark, tuna, *Ichthyosaurus*, and dolphins is the result of convergent evolution that has affected one character or one organ in a similar way in these animals.

The grouping of fishes in taxonomic categories was previously based on the typological criterion, i.e., on the morphological similarity of the specimens to the ideal forms (*eides*) of Plato of which they were considered imperfect copies, or to the original pair of each species, exemplified in the creation narratives of different religions.

The practical problem of identification rests on the difficulty in measuring the degree of similarity between the specimen and its model, often imagined or arbitrarily selected. An interesting case of this difficulty is the double specific name (*Gadus callarias* and *Gadus morhua*) given to the Atlantic cod by Linnaeus in his *Systema naturæ,* in his belief that they were two different species.

From the comparison of the morphological features of fishes a better understanding of the variability of their characters has been attained. Both, morphometric characters (size, colour, etc.) and meristic characters (number of scales, gill rakers, etc.) present a large range of values between two extremes. Therefore it is necessary to work with populations instead of with individual specimens, as was the practice in the past, when some species were determined by single specimens.

This difficulty persists today in systematic work, since in practice, the only criterion applicable is the typological or phenetic one, using morphological, statistical, chromatographic, electrophoretic, and serological methods, to name a few. The only solution, itself not 100 % reliable, is the interbreeding test. Unfortunately, its application is prohibitive because of difficulty and cost. According to one of the most accepted definitions (Mayr, 1969), a species is "a group of actually or potentially interbreeding natural populations which are reproductively isolated from other such groups".

The genetic criterion obviously cannot be applied to taxa superior to the species level (genus, family, and order). In these cases, we return again to the morphological approach. As a consequence, the classifications proposed are numerous and disparate, since they are based on the personal knowledge of the systematist and on his subjective weighing of the characters used.

Although many of the taxonomic names proposed in the past have been rejected, it is useful to know their meaning, since they cast light on the history of Ichthyology and help in evaluating the classical works of previous ichthyologists. Modern classifications examine and weigh all possible characters both individually and as a whole, grouping and organizing the taxa according to an evolutionary criterion. This approach is more valuable since it uses morphological, anatomical, physiological, ecological, paleontological, and embryological data, which are then analyzed by diverse methods.

A new approach, *phylogenetic systematics, cladism* or *Hennigian systematics,* is providing refinements in classification. Cladism relates organisms on the basis of shared advanced, specialized or derived characters. Primitive characters, which may be perpetuated in several lineages, are ignored in establishing relationships. Hennig (1950, 1966) proposed two general new terms to emphasize his approach: *plesiomorphy*, or condition of a character being primitive; and *apomorphy*, condition of a derived or specialized character. These terms are applied to a specific taxonomic level, but always in relation to another group more specialized or more primitive, respectively.

Fish classification has to solve two problems: on the one hand, it must establish the degree of relationship among the living species in a spatial arrangement, so to speak; on the other hand, it should establish the temporal relationship of the species in an evolutionary context. Most modern systems have been organized according to the general lines listed in Table 2.

The following list gives an idea of the variation in the number of taxonomic groups proposed by different systematists for the teleostean ensemble. The number of orders seems to be established at around thirty.

Author and year	Number of orders
Müller (1844)	4
Boulanger (1904)	13
Goodrich (1909)	9
Regan (1909)	28
Jordan (1923)	39
Berg (1940)	41
Bertin and Arambourg (1958)	31
Greenwood *et al.* (1966)	30
Rass and Lindberg (1971)	31
Nelson (1976)	31

The number of fish species surpasses by far the number of species of all the remaining vertebrate groups. The highest number estimated for the existing fish species is 40,000; a conservative estimate of the known fishes to date places the limit between 20,000 and 25,000 species. In contrast, the estimate for mammals is 4,500; birds, 8,500; reptiles 6,000 and for amphibians, 2,500.

The extreme variety of fish forms is due to two main factors: the long evolutionary history of the fish group, which is the oldest in the vertebrate division, and the variety of aquatic habitats with their corresponding diversity of environmental conditions of light, salinity, density, pressure, currents, concentration of oxygen, pH, temperature, nutrients, etc.

Bibl: Compagno (1977); Berg (1958); Cohen (1970); Golvan (1962; Grassé (1958): Jordan and Evermann (1896).

Table 2. Abridged classification of fishes.

Superclass
>   **Agnatha** (fishes without mandibles)

>   *Class*
>>   Cephalaspidomorphi . Lampreys (40 sp.) and hagfishes (30-35 sp.)

Superclass
>   **Gnathostomata**   (fishes with mandibles)
>   *Class*
>>   I. Placodermi   (fossil forms)

>>   II. Chondrichthyes or cartilaginous fishes
>>>   Selachii   sharks (280-300 sp.)
>>>   Batoidei   rays and skates (300 sp.)
>>>   Holocephali   chimaeras (25 sp.)

These two classes (Placodermi and Chondrichthyes) have been sometimes included together as ELASMOBRANCHIOMORPHI (Romer and Parsons, 1978).

>>   III. Osteichthyes or bony fishes
>>   *Subclass*

>>>   Crossopterygii   coelacanths (1 sp.)
>>>   Dipnoi   lungfishes(6 sp.)
>>>   Brachiopterygii   bichirs (11 sp.)
>>>   Actinopterygii   (fishes with fin rays)

>>>   *Infraclass*

>>>   Chondrostei   sturgeons (25 sp.) and
>>>   paddlefishes (35 sp.)
>>>   Holostei   bowfin and gars (8 sp.)
>>>   Teleostei

The most important teleostean orders and families from an archaeological point of view are the following.

>>   *Order*
>>>   *Family*

A. <u>Malacopterygii</u>   Anguilliformes   (some 600 sp.)

>>>   Anguillidae   freshwater eels
>>>   Muraenidae   morays

>>   Clupeiformes (1)   (some 300 sp.)
>>>   Clupeidae   herrings and sardines
>>>   Engraulidae   anchovies

>>   Salmoniformes (1)   (some 525 sp.)

	Salmonidae	salmons and trouts
	Osmeridae	smelts
	Esocidae	pikes
	Cypriniformes (2)	(5000 to 6000 sp.)
	Cyprinidae	carps and minnows
	Catostomidae	suckers
	Siluriformes (2)	(some 2000 sp.)
	Siluridae	North American catfishes

B. Acanthopterygii	Gadiformes	(some 700 sp.)
	Gadidae	codfishes
	Merluciidae	hakes
	Scorpaeniformes	(some 1000 sp.)
	Cottidae	sculpins
	Scorpaenidae	rockfishes
	Perciformes	(some 7000 sp.)
	Serranidae	sea basses
	Centrarchidae	sunfishes
	Percidae	perches
	Sciaenidae	drums
	Labridae	wrasses
	Scaridae	parrotfishes
	Scombridae	tunas, mackerels
	Bleniidae	blennies
	Sparidae	porgies
	Pleuronectiformes	(some 520 sp.)
	Bothidae	lefteye flounders
	Pleuronectidae	righteye flounders
	Soleidae	soles
	Lophiiformes	(some 220 sp.)
	Lophiidae	angler fishes
	Tetraodontiformes	(some 320 sp.)
	Balistidae	triggerfishes

(1) These two orders were part of the old taxon Isospondyli
(2) These two orders formed the old taxon Ostariophysi

The comparative terminology of the fish taxa proposed by several leading taxonomists since Linnaeus, who adopted the classification made by Artedi, is shown in table 3. Notice the absence of the term Fishes (Pisces) as a taxon in the last two classifications.

Table 3. Concordance between several classifications, past and present

Linnaeus 1758	Müller 1844	Jordan 1923	Romer 1933	Berg 1940
----	PISCES Leptocardes Amphioxus	LEPTOCARDII Amphioxus	------	-------
----	--------------	MARSIPOBRANCHII	AGNATHA	PETROMYZONES
		1. Lampreys 2. Myxinidae		MYXINI
----	----	ELASMOBRANCHII	CHONDRICH- THYES	ELASMOBRAN- CHII
		1. Selachii 2. Holocephali		HOLOCEPHALI
PISCES	----	PISCES	OSTEICHTHYES	
1. Apodes	1. Ganoidei	1. Dipneusti	---------	DIPNOI
2. Jugulares	2. Physostomi	2. Crossopterygii	1. Sarcopterygii	TELEOSTOMI
3. Thoracici	3. Physoclisti			Crossopterygii
4. Abdominales	4. Pharyngo- gnathi	3. Actinopterygii	2. Actinopterygii	Actinopterygii
5. Branchiostegi	5. Dipneusti			

75. Clavicle

Fr: clavicule
Ger: Schlüsselbein
Lat: clavicula (pl: claviculæ)
Rus: ключи́ца (pl: ключи́цы)
Sp: clavícula

Paired dermal bone located ventral to the cleithrum in some primitive fishes (*Acipenser, Polyodon,* and *Amia*). It is lost or fused with the true cleithrum in higher teleosts. In primitive actinopterygians, the clavicle expands dorsally into an ascending process which fits into the mesial face of the cleithrum.

76. Cleithrum   *

Fr: cléithrum
Ger: Cleithrum
Lat: os cleithrum (pl: ossa cleithra)
Rus: кле́йтрум
Sp: cleitro

The cleithrum is a bone of dermal origin acting as support for the primary pectoral girdle. It articulates dorsally with the supracleithrum and ventrally with the scapula and coracoid. This bone, belonging to the secondary pectoral girdle, forms the frame of the body wall immediately behind the branchial cavity. Both cleithra meet medially under the heart.

The cleithrum has been used to estimate the age of fishes from the families Cyprinidae, Siluridae, and Esocidae. Casselman (1983) has found that for muskellunge (*Esox masquinongy*), cleithra are more reliable than scales in specimes older than age 10.

Syn: clavicle (Parker, 1868)
Figs: 1 E, 3, 13, 25 B, and 40
Bibl: Rojo (1986); Scott (1977)

77. Condyle

Fr: condyle
Ger: Gelenkknorren; Condylus (= Kondylus)
Lat: condylus (pl: condyli)
Rus: мыщелок
Sp: cóndilo

A round protuberance found sometimes at one end of a long bone, especially when it articulates with another bone.

78. Convergence

Fr: convergence
Ger: Konvergenz
Lat:
Rus: конвергéнция
Sp: convergencia

The similarity of organs or morphological features in two different animal lineages not having common ancestry, but which are subjected to common selective pressures. For example, the hydrodynamic shape and the dorsal fin of lamnid sharks, tunnid fishes, porpoises, and the fossil ichthyosaur reptiles. Another example is seen in plankton feeders, such as the basking shark (*Cetorhinus*) and several species of the herring family (herring, shad), all of which have numerous thin gill rakers.

79. Coracoid   *

Fr: coracoïde
Ger: Rabenschnabelbein; Coracoid
Lat: os coracoideum (pl: ossa coracoidea)
Rus: врановáя (= ворóновидная) кость; коракóид [pl: врановы́е (= ворóновидные) кóсти]
Sp: coracoides

Paired bone of endochondral origin belonging to the pectoral girdle. It occupies the ventral part of the supporting skeleton of the fin and usually has a long forward process. It attaches anteriorly to the cleithrum and supports one or two fin radials. Its upper border usually has a notch that matches a similar one on the scapula, the resulting opening being called the *scapular foramen* .

Figs: 1 E, 3, 13, 25 B, and 38

80. Coracoid cartilage

Fr: cartilage coracoïde
Ger: Coracoidknorpel
Lat: cartilago coracoidea (pl: cartilagines coracoideæ)
Rus: врановой хрящ (pl: врановые хрящи)
Sp: cartílago coracoideo

A cartilaginous bar that makes up the ventral section of the pectoral girdle in chondrichthyans. It has a U shape and supports both pectoral fins.

Syn: coracoid bar

81. Coronoid process

Fr: processus coronoïde
Ger: Coronoidfortsatz
Lat: processus coronoideus (pl: processus coronoidei)
Rus: врановой (= венечный) отросток [pl: врановые (= венечные) отростки]
Sp: proceso coronoideo

1. Name applied to the dorsal branch of the dentary.
2. Process found on the dorsal margin of the angular bone.
3. Vertical expansion of the posterior end of the Meckel's cartilage.

Figs: 17 and 18

82. Coronoids   *

Fr: coronoïdes
Ger: Coronoidea
Lat: ossa coronoidea (sing: os coronoideum)
Rus: венцеподобные (= венечные) кости [sing: в-ная (=в-ная) кость]
Sp: coronoides

Dermal bones bearing teeth located on the upper border of the Meckel's cartilage. There is one pair of coronoids in sturgeons and two in *Amia* and *Lepisosteus*.

Syn: presplenial; splenial (Goodrich); prearticular (Holmgren); intradentary (Regan)

83. Coronomeckelian   *

Fr: coronomeckélien
Ger:
Lat: os coronomeckelium (pl: ossa coronomeckelia)
Rus:
Sp: coronomeckeliano

The small bone formed on the postero-lateral part of the Meckel's cartilage, present in *Amia* and also in the embryos, larvae, and juveniles of many teleostes.

Syn: splenial (Owen); articular sesamoid (Ridewood); supraangular (Holmgren and Stensiö, 1936); os meckeli (Berg, 1940); *d* bone (Bridge, 1977).

Fig: 25 B

84. Cosmoid scales *

Fr: écailles cosmoïdes
Ger: Cosmoidschuppen
Lat:
Rus: космбидные чешӱи (sing: космбидная чешуя)
Sp: escamas cosmoideas

Scales characteristic of fossil placoderms, composed of several layers of hard tissues and covered sometimes with denticles. Depending on the presence or absence of denticles, the cosmoid scales can be classified into denticulate and adenticulate.

Cosmoid scales were first named by Williamson (1849) who discovered them in a Permian crossopterygian and afterwards also found them in numerous fossils of this group. Cosmoid scales are thick, composed of four well defined layers. Outside, there is a thin, external layer of dentine with numerous pores. The second layer is spongy and pierced by winding and branched canaliculi, some of which open to the outside. This layer is made of dentine or a similar substance. Underneath is another spongy layer with osteoblasts, indicating its true bony nature, and finally, a lamellar, proximal layer, made of isopedine, with osteocytes and Sharpey's fibers.

Table 4

85. Cranium

Fr: crâne
Ger: Cranium
Lat: cranium (pl: crania)
Rus: крбниум
Sp: cráneo

See Skull

86. Crest

Fr: crête
Ger: Leiste
Lat: crista (pl: cristæ)
Rus: гребешбк; грбебень
Sp: cresta

Any long and narrow prominence protruding from the surface of a bone.

Figs: 14 and 22

87. Crossopterygians

    Fr: Crossoptérygiens
    Ger: Crossopterygier; Quastenflosser
    Lat: Crossopterygii
    Rus: кистепёрые рыбы
    Sp: Crosopterigios

        This taxonomic group was created and named by Huxley (1861) to include
the fossil actinopterygians having lobate paired fins, i.e., fins that look like tetrapod
limbs.   Crossopterygians are represented nowadays by the only surviving genus
Latimeria, found for the first time in the eastern coasts of South Africa in 1938.
    The most outstanding skeletal features of crossopterygians are:
    a)   the presence of a premaxilla;
    b)   the absence of maxillaries;
    c)   teeth of normal type;
    d)   palatoquadrate not fused to the cranium;
    e)   two dorsal fins;
    f)   radial bones and appendicular musculature extended into the paired fin bases;
    g)   skull composed of two units joined by a moveable articulation;
    h)   cosmoid scales; and
    i)   dermal bones of cosmoid structure.

        The primitive crossopterygians represent the stem from which, according to
the most accepted interpretation, land vertebrates derived.

    Syn   Crossopts
    Tables 2, 3, and 4

88. Ctenii   *

    Fr: spinules
    Ger: Ctenien; Ktenien
    Lat: ctenii (sing: ctenius)
    Rus: шйпики
    Sp: ctenios

        Small expansions recalling the teeth of a comb with their points directed
backwards and found on the exposed area of the ctenoid scales. Their size, distribution,
and number can sometimes be used for taxonomic purposes.

    Syn: spinules
    Fig: 10

89. Ctenoid fishes

    Fr: poissons cténoids
    Ger: Ctenoidfische (= Ktenoidfische)
    Lat: pisces ctenoidei
    Rus: ктенбидные рыбы (sing: ктенбидная рыба)
    Sp: peces ctenoideos

        This taxonomic group, formed by Agassiz (1833), includes those fishes
provided with scales having short pointed spinules (ctenii), such as Pleuronectidae.

90. Ctenoid scales *

Fr: écailles cténoïdes
Ger: Ctenoidschuppen (= Ktenoidenschuppen); Kammschuppen
Lat: squamæ ctenoideæ (sing: squama ctenoidea)
Rus: ктеноидные чешуи (sing: ктеноидная чешуя)
Sp: escamas ctenoideas

Laminar scales of the type called leptoid, derived from the ganoid scales by reduction in thickness of the layer of ganoine. As the name implies, they have pointed expansions called *ctenii* on their posterior field. These spinules make the surface of the fish rough to the touch. They are present in advanced actinopterygians, such as Percidae and Soleidae. Some Pleuronectiformes have both cycloid and ctenoid scales.

Fig: 10
Table 4

91. Cycloid fishes

Fr: poissons cycloïdes
Ger: Zykloidfische (= Cycloidfische)
Lat: pisces cycloidei
Rus: ктеноидные рыбы (sing: ктеноидная рыба)
Sp: peces cicloideos

Under this name, Agassiz (1833) grouped those bony fishes having flexible, overlapping and circular scales which lack ctenii.

92. Cycloid scales *

Fr: écailles cycloïdes
Ger: Zykloidschuppen (= Cycloidschuppen); Rundschuppen
Lat: squamæ cycloideæ (sing: squama cycloidea)
Rus: циклоидные чешуи (sing: цыклоидная чешуя)
Sp: escamas cicloideas

The name cycloid refers to the round shape of these scales. They are arranged quincuncially in rows that completely cover the body and sometimes the nape, cheeks, and operculum. Each scale covers the anterior part of the following scale, and is itself covered partially by the previous one in an imbricated fashion. The surface of these scales is even, lacking expansions and ctenii.
Cycloid scales are of the leptoid or elasmoid type, that is to say, they are flat and thin laminae, resulting from an evolutionary reduction in thickness of the ganoine and isopedine layers of primitive fish scales. They are found most often in malacopterygian fishes. These scales have a center of formation, the focus, around which concentric lines of sclerites are deposited. In some cases, these concentric layers are cut by transverse grooves called *radii* , which can be divided into primary radii, when they start from the focus, and secondary if they are shorter.

Fig: 10
Table 4

93. Cyclospondylous vertebra   *

Fr: vertèbre cyclospondyle
Ger: zyclospondyler (cyclospondyler) Wirbel
Lat: vertebra cyclospondyla (pl: vertebræ cyclospondylæ)
Rus: ЦИКЛОСПОНДЙЛЬНЫЙ ПОЗВОНОК (pl: Ц-ные ПОЗВОНКЙ)
Sp: vértebra ciclospondila

The type of elasmobranch vertebra in which the ring-like calcifications extend only to the chordacentrum or notochordal sheath, while the arches remain cartilaginous. The sevengill shark (*Heptranchias*), the dogfish shark (*Squalidae*), and the Holocephali have cyclospondylous vertebrae.

Fig: 35 A

94. Cyclostomes

Fr: Cyclostomes
Ger: Rundmäuler; Zyclostomen (= Cyclostomen)
Lat: Cyclostomi; Cyclostomata
Rus: круглоробтые рьбы
Sp: Cyclóstomos

Group of fishes established by Duméril (1856) to include lampreys and hagfishes. The name refers to the circular shape of the mouth, which they use to suck the blood and soft tissues of the fishes on which they feed.

See   Agnathans
Table   2 and 4

95. Dental formula

Fr: formule dentaire
Ger: Zahnformel
Lat: formula dentaria
Rus: зубная фбрмула
Sp: fórmula dentaria; fórmula dental

The number of teeth of an individual fish or of a particular species can be expressed by a fraction, with the numerator representing the teeth of the upper mandible and the denominator, those of the lower mandible. This mathematical expression is called *dental formula,* in which the number of teeth on the right and left halves are separated with a hyphen. If there are teeth of different sizes and shapes in the maxillar or in the mandibular symphyses, they are indicated by a number placed between the two sets previously mentioned. The dental formula is more valuable when it represents a sample than when it refers to only one specimen. In the former case, the minimum and the maximum values should also be indicated. The dental formula for the shark *Carcharinus milberti* is

$$\frac{(14 - 16) - 2 - (14 - 16)}{(12 - 15) - 1 - (12 - 15)}$$

where the 2 in the numerator and the 1 in the denominator represent the symphyseal teeth.

When the teeth are very numerous and small, the dental formula is not applicable. In Cypriniformes, the 5th ceratobranchial, also called lower pharyngeal bone, has from one to three rows of teeth. The dental formula for a cyprinid with three rows of pharyngeal teeth, can be expressed as follows:

$$L/5 + 3 + 2 \; : \; R/2 + 3 + 5$$

with the numbers on the left side of the formula representing the outer, middle, and inner number of the left teeth, while the numbers on the right side are written in reverse order.

96. Dentary *

Fr: dentaire
Ger: Dentale
Lat: os dentale (pl: ossa dentalia)
Rus: зубная (= дентальная) кость; [pl: зувные (=дентальные) кости]
Sp: dentario

The dentary is a paired dermal bone, bearing teeth in most bony fishes, that forms the anterior part of the mandible. It has a "V" form, with its apex in an anterior position. Both dentaries meet at the mandibular symphysis.

The dentary is formed by two bony laminae with a central cavity, occupied by a thin, pointed rod of cartilage, known as *Meckel's cartilage* and the anterior process of the angular bone. This bone results, in bony fishes, from the fusion of the dentary proper and one or two splenials. Posteriorly, it has two processes: a dorsal, also called *coronoid process* and a lower or ventral one. The mandibular sensory canal runs along a groove excavated on the outer face of the ventral process.

In Tetraodontiformes (*Tetraodon* and *Diodon*), both dentaries fuse together in the shape of a parrot's beak, with the teeth either lost or represented by small conical prominences.

Syn: dentalo-splenial-mentomandibular (Holmgren and Stensiö, 1936; Pehrson, 1944; Lekander,1949); dento-splenial (Holmgren and Stensiö, 1936; Jollie, 1986); splenial-dentosplenial.

Figs: 1 B, 3, 13, 14 B, and 17
Bibl: Rojo (1986)

97. Dentigerous bones

Fr: os à dents; os dentigères
Ger: zahntragende Knochen
Lat:
Rus: озубленные кости (sing: озубленная кость)
Sp: huesos dentígeros

Dentigerous bones are dermal bones that bear teeth. Teeth are consequently directly related to those bones, either of dermal or endochondral origin, that are associated with the feeding function. These bones are located in the mandibles, the tongue, the mouth, and the branchial apparatus. Teeth are formed independently of the bones but join them in later stages by means of an intermediate basal tooth plate or by connective fibers. The branchial apparatus has endochondral bones, which are never directly associated with the teeth. There is always a dermal bony plate between the endonchondral bone and the teeth.

Here is the list of the bones that bear teeth in one or another species of fishes.

1. Dermal bones associated with the mandibles

> premaxilla
> maxilla
> dentary
> coronoid(s)

2. Dermal bones associated with the buccal cavity

> prevomer (= vomer)
> palatine
> endopterygoid
> ectopterygoid
> parasphenoid
> dermentoglossum

3. The following endochondral bones of the viscerocranium are sometimes associated with dermal dental plates or pads which carry the teeth.

> basihyal (its dermal plate recieves the special name dermentoglossum)
> basibranchials
> anterohyal (= ceratohyal)
> infrapharyngobranchials (= pharyngobranchials)
> epibranchials
> ceratobranchials
> hypobranchials
> basibranchials

Table   1
Bibl: Moss (1972)

98. Dermal bones

Fr: os dermiques
Ger: Hautknochen
Lat:
Rus: покро́вные (=накла́дные) ко́сти [sing: п-ная (=н-ная) кость]
Sp: huesos dérmicos

These bones derive from the dermal mesenchyme of the dermatome, during the embryogenesis of the fish. The classification of bones into dermal and nondermal bones is based on a phylogenetic criterion. Although in the present stage of fish evolution many dermal bones are deeply embedded in the musculature, it is customary to call them dermal bones since they are homologous with the dermal plates of ostracoderms and placoderms. In a general sense, dermal bones are equivalent to membrane bones.

Pehrson (1944) established three types of dermal bones in the skull:

a)   laterosensory canal bones, developed in relation to sensory lines. They represent a very conservative group of bones in the skull useful in comparative studies;

b)   bones derived from the mesenchymous tissues; and

c)   anamestic bones. (See this term).

Dermal bones tend to become fewer as we move up the scale of vertebrates. Primitive fishes have the most. The series of dermal bones that show this trend clearly are the opercular and the orbital series, and the dermal jaw bones.

See   Membrane bones
Syn: achondral bones; covering bones
Table   1
Bibl: Moy-Thomas (1938)

99. Dermal denticles   *

Fr: denticules cutanés
Ger: dermale Dentikel; Hautdentikel; Hautzähnchen
Lat:
Rus: кожные шйпики (sing: кожный шипик)
Sp: dentículos dérmicos

This term is often used to describe the placoid scales because they have a dermal origin and their structure is identical to that of teeth. They are found in modern elasmobranchs, either as scales covering the body of selachians and the head of holocephalans, or as discrete spiny structures on the dorsal surface of batoids.

The term *dermal denticles* seems more anatomically correct than the term *placoid scales*, but the latter is older and more widespread in fish literature and better describes the phylogenetic origin of the placoid scales from the exoskeleton of placoderms.

See   placoid scales
Fig: 8 A
Bibl: Hertwig (1874)

100. Dermarticular

Fr: dermoarticulaire
Ger: Dermarticulare
Lat: os dermoarticulare (pl: ossa dermoarticularia)
Rus:
Sp: dermoarticular

Term proposed by Holmgren and Stensiö (1936) for the small rudimentary bone of dermal origin formed in the mandible of teleosts, that later joins either the angular or the retroarticular bone.

101. Dermis

Fr: derme
Ger: Lederhaut
Lat: dermis
Rus: дерма
Sp: dermis

The dermis is the deepest layer of the skin, separated from the epidermis by an even or slightly ondulated surface. The dermis is thicker than the epidermis and contains the typical cells (fibrocytes) of the connective tissue, fibers of the collagenous and elastic types, blood vessels, nerves, chromatophores, and specialized sensory organs. Scales, and also their derivatives the teeth and the fin rays, are formed in the dermis. A layer, called *subcutis,* very thick in lampreys, separates the dermis from the underlying musculature. The dermis tends to be thinner in scaled fishes than in scaleless ones (Aleev, 1963).

The main function of the dermis is to provide nutrients to the cells of the epidermis by the dialisis process taking place between the blood vessels, very abundant in the dermis, and the outer epithelium, which lacks them.

In some cases the extraordinary proliferation of blood vessels in the dermis provides the fish with a means to carry out the respiratory function for a considerable period of time when the fish might wander onto land, as happens with the eel (*Anguilla*).

Syn: corium

102. Dermocranium

Fr: dermocrâne
Ger: Dermatokranium (= Dermatocranium)
Lat: dermocranium
Rus: вторичный (=кожный) череп
Sp: dermocráneo

The dermocranium is one of the three units that forms the skull of bony fishes. It is the largest unit because of the number of its bones and the area they cover. Histologically, the dermocranium is of dermal origin, since it derives from the connective tissue of the dermis. Phylogenetically, the dermocranium derives from the exoskeleton of the head and pectoral skeleton of armour fishes (ostracoderms and placoderms). During the evolution of fishes, the dermal bones showed a tendency to a reduction in thickness and a migration to deeper areas of the body, thus giving the fish more flexibility. The dermal bones cover the bones of the subjacent units (neurocranium, mandibular and branchial arches), sometimes fusing with them, and eventually, even replacing them completely.

The dermal bones of the dermocranium can be grouped into smaller units, similar to those proposed for the neurocranium and the viscerocranium. In the following list only the most constant bones are given.

Region	Bones	
1. nasal	1.1	nasal
2. orbital	2.0	Circumorbitals bones divided into:
	2.1	supraorbitals
	2.2	infraorbitals
	2.3	suborbitals
3. otic	3.1	intercalar
4. occipital	4.1	tabulars
5. dorsicranial	5.1	frontal
	5.2	parietal
	5.3	pterotic (mixed)

6. basicranial		6.1	parasphenoid
7. oromandibular		7.1	premaxilla
		7.2	maxilla
		7.3	supramaxilla
		7.4	dentary
		7.5	angular (mixed)
		7.6	retroarticular (mixed)
		7.7	coronoid
		7.8	splenial
		7.9	endopterygoid
		7.10	ectopterygoid
		7.11	gular
8. palatine		8.1	prevomer
		8.2	dermopalatine
9. opercular		9.1	preopercle
		9.2	opercle
		9.3	subopercle
		9.4	interopercle

Fig: 14 B
Table 4
Bibl: Parrington (1967)

103. Dermopalatine

Fr: dermo-palatin
Ger: Dermopalatinum
Lat: os dermopalatinum (pl: ossa dermopalatina)
Rus:
Sp: dermopalatino

Allis (1897) gave this name to the paired, dermal bony plate bearing teeth which covers the ventral face of the autopalatine. When present in teleosts, the dermopalatine is fused to the autopalatine, forming one bone named simply *palatine*.

See   Autopalatine and Palatine

104. Dermoskeleton

Fr: squelette dermique
Ger: Hautskelett
Lat:
Rus: кожный скелёт
Sp: dermoesqueleto

The dermoskeleton is the assemblage of bones of dermal origin, resulting from the evolution of the dermal plates of ostracoderms and placoderms. Table 4 shows the evolutionary lines of the main dermal skeletal units in present day fishes.

See   Exoskeleton

Table 4. Evolutionary lines of the dermal plates of Ostracoderms in the larger groups of fishes

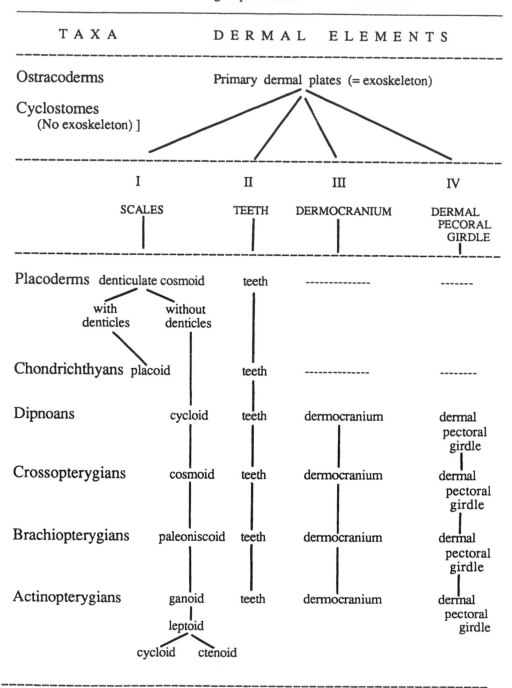

T A X A	D E R M A L E L E M E N T S			
Ostracoderms	Primary dermal plates (= exoskeleton)			
Cyclostomes (No exoskeleton)]				
	I	II	III	IV
	SCALES	TEETH	DERMOCRANIUM	DERMAL PECORAL GIRDLE
Placoderms	denticulate cosmoid → with denticles / without denticles	teeth	-------------	-------
Chondrichthyans	placoid	teeth	-------------	-------
Dipnoans	cycloid	teeth	dermocranium	dermal pectoral girdle
Crossopterygians	cosmoid	teeth	dermocranium	dermal pectoral girdle
Brachiopterygians	paleoniscoid	teeth	dermocranium	dermal pectoral girdle
Actinopterygians	ganoid → leptoid → cycloid / ctenoid	teeth	dermocranium	dermal pectoral girdle

105. Dermosphenotic   *

Fr: dermosphénotique
Ger: Dermosphenoticum
Lat: os dermosphenoticum (pl: ossa dermosphenotica)
Rus: дермосфенотикум
Sp: dermoesfenótico

Name given by Parker to the last bone (IO₆) of the infraorbital series. The dermosphenotic is located on the upper posterior part of the orbit, being characterized by a bifurcation of the infraorbital sensory canal that joins the supraorbital canal and the lateral line canal.
In many Teleosts, it acts as a hinge for the articulation of the skull with the circumorbital ring. In spite of its name, it has no relation to the otic capsules. It has erroneously been called postfrontal, intertemporal, and even sphenotic.

Figs: 1 A and 14 B

106. Dermosupraoccipital

Fr: dermo-supraoccipital
Ger: Dermosupraoccipitale
Lat: os dermosupraoccipitale (pl: ossa dermosupraoccipitalia)
Rus:
Sp: dermosupraoccipital

Dermal paired bone that usually joins the endochondral supraoccipital. In Siluridae, it extends posteriorly in a toothed process that secures the nuchal disc.

Syn: parietooccipital; postparietal (Berg,1940); dermal supraoccipital

107. Dermotrichs

Fr: dermotriches
Ger: Dermatotrichen (sing: Dermatotrich)
Lat: dermotrichia (sing: dermotrichium)
Rus: дермотрихии (sing: дермотрихия)
Sp: dermotricos; dermatotricos

Term used by Goodrich (1904) to designate the fin rays, since all are of dermal origin. The dermotrichs comprise all simple, segmented, either keratinized or bony rays of fish fins.
They are divided into four groups: *ceratotrichs*, found in cartilaginous fishes; *actinotrichs* , found in both cartilaginous and bony fishes; *lepidotrichs* , present only in bony fishes; and finally *camptotrichs*, typical of dipnoans and crossopterygians. By extension, the spiny rays of acanthopterygians are called *acanthotrichs*.

See the terms in italics.
Syn: dermatotrichs

108. Diplospondylous vertebra *

Fr: vertèbre diplospondyle
Ger: diplospondyler Wirbel

Lat: vertebra diplospondyla (pl: vertebræ diplospondylæ)
Rus: ДИПЛОСПОНДИЛНА̄ЛЬНЫЙ ПОЗВОНО̄К (pl: Д-ные ПОЗВОНКЙ)
Sp: vértebra diplospondila

Type of vertebra having two centra: a *precentrum*, which lacks neural and hemal arches; and one *postcentrum*, that possesses them. This type of vertebrae are present in the caudal region of *Amia*. This phenomenon is known as *diplospondyly*.

109. Dipnoans

Fr: Dipneustes
Ger: Dipnoer; Lungenfische
Lat: Dipnoi; Dipneusti
Rus: ДВОЯКОДЫ̄шащие РЫ́бы
Sp: Dipnoos

Dipnoans, a subclass of osteichthyan fishes, are characterized by the presence of one or two functional lungs, in addition to gills. The lungs are used to breath atmospheric air during the dry season while the gills function when the fish is in the water. Their most notable osteological features are:
a)   the lack of premaxilla and maxilla;
b)   the presence of three pairs of dental laminae;
c)   the fusion of the palatoquadrate bone to the skull;
d)   the presence of one dorsal fin;
e)   the extension of the radials and fin muscles into the base of the paired fins; and
f)   body covered with cosmoid scales.

Three genera of dipnoans are present today. In Australia, *Neoceratodus,* which has only the right lung, is found in some rivers of Queensland. *Protopterus* in Africa and *Lepidosiren* from the Amazon Basin of South America, both with two lungs.

Syn: lungfishes
Tables:
2, 3, and 4
Bibl: Lison (1941)

110. Dorsal fin

Fr: nageoire dorsale
Ger: Rückenflosse; Dorsalflosse
Lat: pinna dorsalis (pl: pinnæ dorsales); notopterygium (pl: notopterygia)
Rus: СПИННО̄Й ПЛАВНЙК
Sp: aleta dorsal

This fin (notopterygium) is located on the back of most fishes including the primitive cyclostomes. The main function of the dorsal fin is to maintain the equilibrium of the fish by avoiding rolling. In fishes with a long dorsal fin (*Amia* and *Anguilla*), it also aids in locomotion. Sometimes the fin is divided into two or more fins, as in the Gadidae family. In acanthopterygians the dorsal fin is split into two sections : one anterior, with spiny rays and the other, posterior, with soft rays.
The presence of spines (modified soft fin rays) in front of the dorsal fin, is rather frequent, as in dogfish, Ictaluridae and sticklebacks. Among the most outstanding

modifications of the dorsal fin, we can mention the adhesive disc of the remora and the flexible barbel of the angler fishes.

In several families, for example, in *Polypterus* and in Scombridae, the dorsal fin splits into a variable number of small fins, known as *finlets* or *pinnulae*.

See   Illicium
Figs: 6 and 13
Bibl: Emelianov (1973); François (1958); Goodrich (1906); Lindsey (1955).

111. Dorsicranium

Fr: dorsicrâne
Ger: Dorsicranium; Schädeldach
Lat: dorsicranium (pl: dorsicrania)
Rus: дорсикра́ниум
Sp: dorsicráneo

A rather vague term used to include, in a general way, the endochondral and membrane bones of the dorsal region of the skull.

112. Dyssospondylous vertebra   *

Fr: vertèbre dyssospondyle
Ger: dyssospondyler Wirbel
Lat: vertebra dyssospondyla (pl: vertebræ dyssospondilæ)
Rus: диссоспондила́льный позвоно́к (pl: д-ные позвонки́)
Sp: vértebra disospondila

It is the vertebra whose components fail to join into one unit during its embryology, as in sturgeons, gars, and *Amia* .

113. Ectocoracoid

Fr: ectocoracoïde
Ger: Ectocoracoid
Lat: os ectocoracoideum (pl: ossa ectocoracoidea)
Rus: эктокорако́ид
Sp: ectocoracoides

Name given by Regan (1909) to the paired bone located at the base of the pectoral girdle in sticklebacks (*Gasterosteus*).

114. Ectopterygoid   *

Fr: ectoptérygoïde
Ger: Ectopterygoid
Lat: os ectopterygoideum (pl: ossa ectopterygoidea)
Rus: нару́жная крыловѝдная кость; эктоптериго́ид
Sp: ectopterigoides

Paired bone of dermal origin that occupies the posterior part of the palatine arch, sometimes called simply *pterygoid* , especially when there is no endopterygoid.

The ectopterygoid forms part of the palate in teleostean fishes. It articulates anteriorly with the palatine, posteriorly with the quadrate, and mesially with the endopterygoid (if present).

In *Amia* and other primitive actinopterygians, the ectopterygoid has one or two rows of teeth.

Syn: pterygoid
Figs: 1 D, 3, 13, 25 A, and 28

115. Elasmobranchs

Fr: Elasmobranches
Ger: Elasmobranchier; Plattenkiemer
Lat: Elasmobranchii
Rus: пластинчатожаберные рыбы
Sp: Elasmobranquios

Taxonomic group created by Bonaparte (1832) to include sharks, rays, and skates. Its name refers to the laminar structure of their gill filaments. The class Elasmobranchii, comprising nowadays some 700 to 800 mostly marine species, can be subdivided into two major subgroups: the selachians (sharks) and the batoids (rays and skates).

The most important skeletal features common to the group are, the presence of
a)   a cartilaginous skeleton;
b)   five to seven independent gill arches;
c)   placoid scales; and
d)   a copulatory organ in males of some species, called *myxopterygium* or *clasper*.

The cartilaginous pieces of the skeleton of elasmobranchs are not usually preserved, but teeth, so characteristic of these predators, are a good identifying element in archaeological sites, since they are always preserved thanks to their enamel or enameloid component. Placoid scales or denticles are also a good identification tool for the archaeologist, but because of their small size, special care should be taken to collect denticles in the dig. Vertebrae and spines have also been found in archaeological sites of recent age.

Tables 2 and 3
Bibl: Compagno (1973, 1977); Daniel (1934); Gilbert and Mathewson and Rall (1967); Moss (1977); Reif (1978); Springer and Garrick (1964); Thompson and Springer (1965).

116. Endopterygoid   *

Fr: endoptérygoïde
Ger: Endopterygoid
Lat: os entopterygoideum (pl: ossa endopterygoidea)
Rus: энтоптеригоид
Sp: endopterigoides

Name proposed by Goodrich (1930) for the paired dermal bone formed in the middle part of the pterygopalatine arch. The endopterygoid articulates with the palatine and the ectopterygoid. It bears teeth in the primitive teleosts (Albulidae), but lacks them in the most advanced teleosts.

Syn: entopterygoid (Berg, 1940); Gregory, (1951); mesopterygoid
Figs: 1 D, 3, 13, 25 A, and 28.

117. Endoskeleton

Fr: endoskelette
Ger: Endoskelett; Innenskelett
Lat:
Rus: вну́тренний скеле́т
Sp: endoesqueleto

The endoskeleton is made up of all the bones formed originally deep in the fish body. These bones constitute the *axial skeleton* (neurocranium, viscerocranium and vertebral column) and the *appendicular skeleton* with its three divisions (the primary pectoral girdle; the pelvic girdle; and the supports for the radii of both paired and median fins).

118. Epibaseost

Fr: épibaséoste
Ger: distales Segment des Flossenträger
Lat:
Rus:
Sp: epibaseosto

It is the distal element of the pterygiophore, which articulates directly with the rays of the dorsal and anal fins.

See   Pterygiophore
Syn: baseost
Fig: 41

119. Epibranchials *

Fr: épibranchiaux
Ger: Epibranchialia
Lat: 1. cartilagines epibranchiales (sing: cartilago epibranchialis)
       2. ossa epibranchialia (sing: os epibranchiale)
Rus: 1. эпибранхиа́льные (=ве́рхнежа́берные) хрящи́
       2. эпибранхиа́льные (=ве́рхнежа́берные) ко́сти
Sp: epibranquiales

The epibranchials, cartilaginous in elasmobranchs and bony in osteichthyans, are arranged in pairs in the branchial apparatus. They articulate dorsally with the pharyngobranchial and ventrally with the ceratobranchial bones in teleosts but they are missing in the last branchial arch (V). Very often the epibranchials are covered with toothed pads arranged regularly on their inner and outer sides. The epibranchials are lost in some teleosts.

Figs: 4 and 25 A

120. Epicentrals *

Fr: épicentraux (sing: épicentral)

Ger: Epicentralia
Lat: ossa epicentralia (sing: os epicentrale)
Rus:
Sp: epicentrales

They are short and rod-like bones attached to the centra of the most anterior vertebrae. *Merluccius* has three or four pairs of epicentrals.

121. Epidermis

Fr: épiderme
Ger: Oberhaut
Lat: epidermis
Rus: эпидéрмис
Sp: epidermis

The epidermis, formed at the expense of the embryonic ectoderm, forms the outermost layer of the fish skin, and is very similar in structure to the human buccal mucosa. The epidermis constitutes a stratified epithelium made up of several layers of cells, from 4 to 6 in selachians, and from 10 to 30 in teleosts. This number varies not only with the species, but also with the body region and the age of the fish. In *Osmerus,* for example, the number of layers varies between 12 and 15 on the head and only from 4 to 5 in the fins.

The epidermis is formed of living cells cemented together by an intercellular matrix. The deepest layer, the *stratum germinativum,* has columnar cells, which with their continuous duplication, compensate for the loss of the outermost cells, which are older and exposed to constant wear. The epidermal cells are interspersed with mucous cells and sometimes with poisonous glands.

The epidermis of cyclostomes is covered by a thin acellular cuticle, secreted by the epidermal cells. This cuticle forms a true *stratum corneum.* A similar situation occurs in the pearl organs of some teleosts.

122. Epihyal   *

Fr: épihyal
Ger: Epihyale
Lat: os epihyale (pl: ossa epihyalia)
Rus: эпихиáльная кость (pl: эпихáльные кóсти)
Sp: epihial

Paired bone of endochondral origin, that joins the hyomandibular and the symplectic through the interhyal. It articulates with the ceratohyal by a suture in some families (Gadidae) or by a thin layer of cartilage in most bony fishes.

Since the modern interpretation considers this bone as the dorsal ossification of the ceratohyal, the name epihyal is replaced in modern works by *dorsal ceratohyal* or *posterohyal.*

Syn: posterohyal; dorsal ceratohyal
See   Ceratohyal
Figs: 1 F, 3, 25 B, and 32

123. Epioccipital

Fr: épioccipital
Ger: Epioccipitale

Lat: os epioccipitale (pl: ossa epioccipitalia)
Rus:
Sp: epioccipital

See   Epiotic

124. Epiotic   *

Fr: épiotique
Ger: Epioticum
Lat: os epioticum (pl: ossa epiotica)
Rus: вёрхнеушнáя кость (pl: вёрхнеушнь́е кóсти)
Sp: epiótico

Name proposed by Huxley (1858) for the dorsal ossification of the otic capsule in actinopterygians. The epiotic covers the posterior semicircular canal. Recently Patterson (1975) has shown that the epiotic is an ossification of the occipital arch which has invaded the otic region. In consequence, he proposes the term *epioccipital* to replace the better known *epiotic* . The name epioccipital is appearing more and more often in recent fish literature.

Syn: epioccipital
Figs: 2, 13, and 14 A

125. Epurals *

Fr: épuraux (sing: épural)
Ger: Epuralia
Lat: ossa epuralia (sing: os epurale)
Rus: епурáлии; епурáльные кóсти (sing: епурáльная кость)
Sp: epurales

The median bones lying behind the last neural spine, whose function is to support the upper caudal fin rays. They are located above the urostyle and their number varies from three, in primitive actinopterygians, to one, in more advanced fishes.
The term "epural" was proposed by Huxley (1859) to mean "those isolated ray bearing bones situated immediately dorsal to the urostyle in *Gasterosteus* ."

See   Caudal skeleton
Figs: 13 and 37

126. Ethmoid   *

Fr: ethmoïde
Ger: Ethmoid
Lat: os ethmoidale (= ethmoideum); [pl: ossa ethmoidalia (= ethmoidea)]
Rus: обонЯтельная (= решётчатая) кость; [pl: о-ные (=р-ые
        кóсти)]; этмóид]
Sp: etmoides

Median bone of perichondral origin formed in the nasal septum between the two nasal capsules. It fails to ossify in some teleosts, remaining as a cartilage during the adult stages of the fish.

Syn: hypethmoid; dermethmoid
Figs: 2, 13, and 14 A

127. Exoccipital   *

Fr: occipital latéral
Ger: Occipitale laterale; Exoccipitale
Lat: os exoccipitale (pl: ossa exoccipitalia)
Rus: боковáя затьɪлочная (= экзокципитáльная) кость
Sp: occipital lateral

Paired bone of endochondral origin located on each side of the foramen magnum, which it frames laterally. In many fishes, it carries an articular facet that joins it with the first vertebra.

Syn: lateral occipital
Figs: 2 and 14 A

128. Exoskeleton

Fr: exosquelette
Ger: Aussenskelett; Exoskelett; Ektoskelett
Lat:
Rus: нарýжный скелéт
Sp: exoesqueleto

The exoskeleton comprises all the bony elements that originally were developed in the fish dermis, from the preexisting connective tissue of the skin. The bones originating from the exoskeleton, called *membrane* or *dermal bones,* are located in the head and in the pectoral girdle. Scales and their derivatives are also considered part of the exoskeleton.

In bony fishes (Osteichthyes), many of these bones migrated to deeper layers of the body, well below the dermis, forming the so-called *secondary exoskeleton.*

Table 4

129. Face

Fr: face
Ger: Fläche; Ansicht
Lat: facies (pl: facies)
Rus: лицó (pl: лицá)
Sp: cara

Term applied to any bone surface.

130. Facet

Fr: facette
Ger: Facette
Lat:
Rus: фасéтка (pl= фасéтки)
Sp: faceta

A facet is a flat or slightly curved surface in a bone, where it generally articulates with another bone.

131. Fins

Fr: nageoires
Ger: Flossen (sing: Flosse)
Lat: pinnæ (sing: pinna)
Rus: плавники́ (sing: плавни́к)
Sp: aletas

The membranous expansions supported by rays found on the middle line and on the sides of the fish body. Their scientific name (*pterygium* [pl: *pterygia*]), reflects their similarity in shape and movement to the avian wing. The position of the fins on the body is regulated by hydrodynamic principles based upon different body shapes and swimming speed.

In cartilaginous fishes, the fins are thick expansions of the body, supported by numerous basal cartilages extended well into the fin and by dermic rays (ceratotrichs) covered by the skin. The ceratotrichia provide better support when they reach the border of the pectoral fin, in which case, the fin is labelled "plesodic". When they end short of it, the label "aplesodic" is applied.

In actinopterygians, the fins are supported by a small number of basal bones called *radials, actinosts,* or *pterygiophores,* to which the fin rays are attached. The fins rays are covered by a thin membrane formed mainly by the epidermis.

In crossopterygians and some dipnoans, the fins have a fleshy base reinforced by a well-developed skeleton to which the rays are affixed.

Fins are classified according to their position into two types: median fins, when they are implanted on the median line or plane of the body; and paired fins, when they are set in pairs on the sides of the body. The dorsal fin (*notopterygium*), is found on the back of the fish, and has as its main function the provision of stability, i.e., to prevent the fish body from rolling. Sometimes, especially in the case of fish with elongated dorsal fins, propulsion can be helped by undulations of the dorsal fin. In the posterior end of the body, the caudal fin (*uropterygium*) provides propulsion and direction in a combination of propellor and rudder functions. On the middle ventral side of the fish, the anal fin (*proctopterygium*) also contributes to stability, and occasionally to movement, in combination with the dorsal fin. These median fins are the oldest in the evolutionary history of fishes. Sometimes, one or all of them have been lost in some specialized forms.

Although different in shape and structure, the adipose fin is sometimes included with the median fins. It is characteristic of some primitive teleosts, such as, Salmonidae, Siluridae, and Osmeridae.

Paired fins are also classified according to position: pectoral fins (*omopterygium*) set out on the body at the level of the thorax, and pelvic fins (*ischiopterygium*) originally located back in the abdominal region. The position of the pelvic fins varies from a posterior attachment to a thoracic, jugular, or even mandibular position. In some instances, they disappear as in eels. The paired fins provide hydrostatic stability, making it possible to change direction, brake, and perhaps propel, acting at the same time as keel, rudder, brakes, and propellor.

Syn: ichthyopterygium
Figs: 6, 13, and 37

132. Fish evolution

Fr: évolution des Poissons
Ger: Evolution der Fische
Lat: evolutio piscium
Rus: эволюция рыб
Sp: evolución de los Peces

From our present knowledge of the fish record, the evolution of fishes can be traced to the beginning of the Paleozoic. By the end of the Paleozoic era there were already in existence forms very similar to present ones. The future of fish evolution was largely determined within the Paleozoic.

By Devonian times the two larger divisions of agnathans and gnathostomes already existed and their several branches were well established, so that at the beginning of this period even the advanced fishes were well defined.

The vast majority of agnathans (Osteostraci, Anaspida, Heterostraci, and Thelodonti) became extinct at the end of the Devonian. Only lampreys and hagfishes have survived to the present, with a scarce representation of species.

Gnathostomes had better success. Of the six divisions into which they can be divided (acanthodians, actinopterygians, dipnoans, crossopterygians, placoderms, and chondrichthyans), only two have disappeared: the placoderms, at the beginning of the Carboniferous and the acanthodians, which terminated in the Permian. The remaining groups, with various success, blossomed into a rich variety of forms, so that, within the phylum Chordata, and more precisely, in the group vertebrates, fishes make up the most numerous taxon as far as the number of species is concerned (20,000 to 40,000 species).

The most successful group of gnathostomes is that of the actinopterygians. Within this group, three levels of organization can be readily established, and although the names applied to them are not accepted in modern classifications, they still have a didactic and pragmatic value.

1st level. *The chondrostean level*

The fishes from this division, the most primitive of the group, have developed from the Devonian to the Permian. Sturgeons (*Acipenser*) and paddlefishes (*Polyodon* and *Psephurus*) are their more modern representatives.

2nd level. *The holostean level*

This group, which appeared in the Triassic, is represented in North American waters by the gars (*Lepisosteus*) and bowfin (*Amia*).

3rd level. *The teleostean level*

The Teleosteans, formed by the more advanced fishes, colonized practically all marine and freshwater habitats. It comprises all those actinopterygians which began to evolve in the Cretaceous, and which culminate today in a variety of forms never equalled by any other vertebrate group.

Tables 2 and 3
Bibl: Briggs (1966); Gosline (1965); Gregory (1951); Myers (1958); Woodward (1942)

133. Fishes

Fr: Poissons
Ger: Fische (sing: Fisch)
Lat: Pisces (sing: piscis)
Rus: Рыбы (sing: рыба)
Sp: Peces (sing: pez)

Fishes are aquatic vertebrates which breathe by means of gills (branchiae), although a group of fishes, the lungfishes (Dipnoi) can also breathe with the aid of lungs. All are poikilothermous, that is, they lack a mechanism for maintaining a constant body temperature, which varies in fishes within a wide range according to the surrounding environment, except in some sharks and tunas, in which body temperature can be as much as 10 º C higher than the surrounding water.

Two features almost universal in fishes are the presence of scales, derived from the exoskeleton of fossil ostracoderms and placoderms, and the presence of fins supported by dermic rays.

It is worth noting that not all of the above-mentioned characters are present in all fishes and that they are not exclusive to this vertebrate group. For this reason, PISCES, as a taxonomic group, has been eliminated in modern classifications (see Table 3). The name PISCES (FISHES) is preserved because of its historical and didactic value.

Rostlund (1952) states that "of the six hundred or so species of freshwater fishes in the United States, only one hundred or so were the object of aboriginal fishing".

Tables 2 and 3
Bibl: Atz (1968,1969); Bailey *et al.* (1970); Balon (1980); Berg (1940); Bond (1979); Caillet, *et al.* (1986); Carlander (1969, 1977); Clemens and Wilby (1949); Cohen (1970); Darlington (1957); Dean (1916, 1917,1923); Demolle, Maier and Wundsch (1964); Hubbs and Lagler (1958); Hureau and Monod (1978); Lagler, Bardach, Miller and Passino (1977); Leim and Scott (1966); Moyle and Cech (1982); Muus (1971); Norman (1958); Scott and Crossman (1973).

134. Fissure

Fr: fissure
Ger: Spalte
Lat: fissura (pl: fissuræ)
Rus: борозда́ (pl: бо́росды); щель (pl: ще́ли); извӣлина
        (pl: извӣлины)
Sp: fisura

This name applies to any furrow, notch or stria naturally present in a bone.

135. Fontanelle

Fr: fontanelle
Ger: Fontanelle
Lat: fontanella (pl: fontanellæ)
Rus: родничо́к (pl: родничкӣ); фонтане́ль (pl: фонтане́ли)
Sp: fontanela

Any opening in the skull due to a lack of ossification of the preexisting cartilaginous or connective tissues. The fontanelles are located between two or several adjacent bones, and during the life of the fish, they are closed by membranes

Syn: fontanel

136. Foramen

Fr: foramen
Ger: Öffnung; Loch
Lat: foramen (pl: foramina)
Rus: отве́рстие (pl: отве́рстия)
Sp: foramen

Any natural opening of variable size present in bones to allow the passage of nerves and blood vessels.

Figs: 17 and 32

137. Foramen magnum

Fr: foramen magnum
Ger: Foramen magnum; Hinterhauptsloch
Lat: foramen magnum
Rus: большбе отве́рстие
Sp: foramen magno

The foramen magnum, the largest of the cranial foramina, is located in the occipital region of the skull. Through this foramen, the spinal cord emerges from the skull. It is formed, in most cases, by four bones: the supraoccipital, dorsally; the two exoccipitals, one on each side, and ventrally, the basioccipital.

138. Fork length

Fr: longueur à la fourche
Ger: Gabellänge
Lat:
Rus: длина́ до хвостовбй развйлки
Sp: longitud de la furca; longitud furcal

Distance, in a straight line, between the tip of the snout and the end of the median rays of the tail. This measurement, not used in systematic studies, is very much utilized in fishery biology.

Fig: 7

139. Fosse

Fr: fosse
Ger: Grube
Lat: fossa (pl: fossæ)
Rus: я́мка (pl: я́мки)
Sp: fosa
A cavity in the surface of a bone into which fits the process of another bone.

140. Frontal *

Fr: frontal (pl: frontaux)
Ger: Frontale
Lat: os frontale (pl: ossa frontalia)
Rus: лóбная (= фронтáльная) кость [pl: л-ные (= ф-ные) кóсти]
Sp: .frontal

Paired bone of dermal origin that encloses the supraorbitary sensory canal. The frontals cover a large area of the neurocranium, occupying part of the ethmoidal, orbitosphenoid, and otic regions. They meet along the middle line, sometimes fusing with the lateral ethmoids. In the orbital area, they overhang the orbits articulating with some of the circumorbital bones, the sphenotic, and the pterotic. Posteriorly they meet the parietals. In some cases, as in Gadidae, both frontals fuse into a single bone.

In the *parietolateral type* of skull, the parietals are separated allowing the frontals to suture with the supraoccipital. Frontals are present in all chondrosteans, holosteans, crossopterygians, brachiopterygians, and teleosteans.

It is agreed by most fish embryologists that the frontal of actinopterygians corresponds to the parietal of tetrapods, but because this name is already well established in fish literature, there is tacit consensus to keep it.

Figs: 2, 13, and 14 A

141. Fulcral scales *

Fr: fulcres
Ger: Fulkren; Fulcren
Lat: fulcra (sing: fulcrum)
Rus: фýлькры (синг: фýлькра); вѝльчатые чешýи
(sing: вѝлчатая чешуя̀)
Sp: escamas fulcrales; fulcros

The unsegmented, non-branched rays present in front of the anterior dorsal, anal and caudal fins of chondrosteans (sturgeons) and holostean (*Amia*) fishes. Fulcral scales are also present in some teleosts.

Syn: fulcres
Fig: 9 D

142. Ganoid fishes

Fr: poissons ganoïdes
Ger: Ganoidfische
Lat: pisces ganoidei
Rus: ганóидные рыбы (sing: ганóидная рыба)
Sp: peces ganoideos

A taxonomic group established by Agassiz (1833) for those fishes covered with ganoid scales: *Acipenser, Polypterus,* and *Lepisosteus*.

Table 3

143. Ganoid scales

Fr: écailles ganoïdes

Ger: Ganoidschuppen
Lat: squamæ ganoideæ (sing: squama ganoidea)
Rus: ганбидные чешуй (sing: ганбидная чешуя)
Sp: escamas ganoideas

The ganoid scales are formed by a bony layer of cosmine sandwiched between an external layer of ganoine, which gives the scale its characteristic shine and hardness, and an internal layer of isopedine. Their shape is slightly rhomboid. The ganoid scales cover the fish body overlapping at their borders, giving the fish a hard rigid armour.

Ganoid scales are found in the most primitive actinopterygians, (sturgeons, *Lepisosteus,* and *Amia*), although in these last two species, the layer of ganoine is vestigial. The ganoid scales evolved from cosmoid scales by atrophy of the layers of cosmine and the spongy tissue. They are formed in the embryo from a laminar placode located in the dermis, over which connective tissue cells secrete a layer of ganoine and below it, the osteoblasts deposit a layer of isopedine.

Fig: 8 B
Table   4

144. Gill rakers

Fr: branchiospines
Ger: Siebfortsätze; Kiemenreusendorne
Lat: branchiospinæ (sing: branchiospina)
Rus: жáберные тычйнки (sing: жáберная тычйнки)
Sp: branquispinas

The branchial arches have, in their inner surface, one or two rows of prominent expansions of variable size, known as *gill rakers*. Gill rakers have an internal cartilaginous or bony rod covered by the pharyngeal mucosa. The covering epithelium is provided with asperities or denticles that make gill rakers an efficient device for retaining the small food particles carried by the water that crosses the gills.

There is an intimate relationship between the size and number of gill rakers and the fish diet. Plankton feeders have long, narrow, and densely packed gill rakers, that are an efficient device for filtering water while retaining small food particles. Predator fishes have less numerous and more widely spaced gill rakers, their number and size being inversely related to the size of their prey.

The number of gill rakers is used as an important meristic character to separate sibling species and geographical populations. It is customary to count only the gill rakers of the first left branchial arch, giving separately the number for those on the upper and lower arm of the arch, as well as the total value. It is also important to give the minimum, maximum, and mean value when dealing with a sample.

See   Branchicteniae

145. Gills

Fr: branchies
Ger: Kiemen
Lat: branchiæ
Rus: жáбры (sing: жáбра)
Sp: branquias

The gills or branchiae are the typical respiratory organs in fishes. Each one consists of a cartilaginous or bony skeleton surrounded by muscular tissue, respiratory epithelium, and blood vessels. This assemblage is called a *branchial arch*. The number of branchial arches varies from 16 in some cyclostomes to 4 in teleosts.

In cyclostomes and cartilaginous fishes, the gill skeleton is cartilaginous and therefore will rarely be found in archaeological sites. Except in cyclostomes, a branchial arch consists typically of five pairs of elements joined at the bottom to a row of median elements. The paired cartilages or bones are named, from top to bottom: pharyngobranchial, epibranchial, ceratobranchial, and hypobranchial. The middle series consists, from front to back, of the basihyal and two or three basibranchials or copulae (See these terms). This branchial skeleton is of endochondral origin.

Some of the branchial bones can easily be identified, especially the long ones and those with dermal tooth plates (pharyngobranchials). The fifth ceratobranchial of ostariophysans (Cypriniformes and Siluriformes), frequently called, although not very appropriately, *lower pharyngeal*, has a special shape which could make species identification possible. Its dorsal face bears a dermal plate with strong teeth.

In most bony fishes, the bones of the branchial apparatus are usually associated with teeth implanted on pads of different size and shape. These pads are bony plates of dermal origin.

146. Glossohyal

Fr: glossohyal
Ger: Glossohyale
Lat: os glossohyale (pl: ossa glossohyalia)
Rus: языкоглоточная кость (pl: языкоглоточные кости)
Sp: glosohial

Bony plate of dermal origin, to which teeth are sometimes attached, and which covers dorsally the cartilaginous or bony basihyal in some fishes (Salmonidae).

See   Basihyal
Syn: glossohyal (Ridewood, 1904; Chapman, 1941); entogossum; dermal basihyal.
Fig: 34 A

147. Gnathostomes

Fr: Gnathostomes
Ger: kiefermäulige Fische; Kiefermünder
Lat: Gnathostomata
Rus: челюстные (=челюстноротые) рыбы [sing: челюстая (=челюстноротая) рыбы
Sp: Gnatóstomos

Group of fishes opposed to the agnathans, so designated to include fishes with mandibles and all other anatomical elements related to them. The gnathostomous division encompasses all extant fishes, except the lampreys and hagfishes.

Table   2

148. Gonapophysis

Fr: gonapophysis

Ger: Gonapophyse
Lat: gonapophysis (pl: gonapophyses)
Rus: гонапофиз
Sp: gonapófisis

The enlarged hemal arches that articulate with the gonopodium in some teleost families (Poeciliidae and Horachthyidae)

149. Gonopodium

Fr: gonopode
Ger: Gonopodium
Lat: gonopodium (pl: gonopodia)
Rus: gonopōdij
Sp: gonopodio

The males of live-bearing species of Poeciliidae have a copulatory organ called *gonopodium,* that results from the modification of the rays and the skeleton of the anal fin. The 3rd, 4th, and 5th anal rays lengthen caudally and become completely moveable. The gonopodium is supported by strengthened and enlarged anal pterygiophores.

The contraction of a set of muscles in front of the gonopodium moves it forward, making it possible the deposition of sperm near the female genital pores. At the same time a few hemacanths or hemal spines transmit the movements of the vertebral column to the pterygiophores by means of muscles and ligaments. The anal rays are either modified in the shape of a trough or joined into a tube for the passage of the sperm.

In the Horachthyidae, the anal rays are modified into a more complex gonopodium. Since this family of teleosts is oviparous, it ensues that there is no relationship between the presence of a gonopodium and the viviparous condition; this conclusion is supported by the presence of claspers in oviparous elasmobranchs.

150. Gular plate   *

Fr: plaque gulaire
Ger: Gularplatte
Lat: os gulare (ossa gularia)
Rus: гулярная пластинка (pl: гулярные пластинки)
Sp: placa gular

Dermal bones located at the base of the throat of primitive actinopterygians, extending backwards from the mandibular symphysis, and arranged in one or three rows. Because of their laminar shape they are also called *gular plates.*

There is only one median gular plate bordered by two rows of branchiostegals in *Amia* and in some members of the North American families Elopidae, Megalopidae, and Albulidae. The median plate was called *intergular* by Hubbs (1919). In some lungfishes there is a second, posterior median plate.

In *Latimeria, Polypterus,* and *Calamoichthys,* there are two lateral gulars located between the median plate and the rows of branchiostegal rays (when all are present). In some lungfishes, there may be two pairs of lateral gulars, one anterior, and the other posterior.

Fig: 14 B

151. Head

Fr: tête
Ger: Kopf
Lat: caput (pl: capita)
Rus: 1. голова́ (pl: головы́) 2. голо́вка
Sp: cabeza

1. Since fishes lack a well-defined neck, in contrast to the remaining vertebrates, it is difficult to establish the posterior limit of the head. By consensus, it has been accepted that the division between the head and the trunk is marked by the posterior border of either the opercular bone or the membrane covering it.

The morphology and anatomy of the head, especially of its bones and muscles, are essential to the study of fish systematics and evolution, because of the predominant role played by the head in the most important activities of the fish: feeding, defense, attack,and even locomotion.

When the cephalic bones found in an archaeological site are associated with postcranial bones, it is assumed that the fish has been consumed *in situ*. On the other hand, when the cranial bones are absent, this usually indicates the fish was beheaded and cleaned before being transported to the site or exchanged in commercial activities.

2. Any round process in a bone

Fig: 7

152. Head length

Fr: longueur de la tête
Ger: Kopflänge
Lat: longitudo capitis
Rus: дли́на́ головы́
Sp: longitud de la cabeza

The horizontal distance between the projections of the tip of the snout and that of the bony or membraneous border of the opercular membrane. The length of the head, as a ratio of the total fish length, is commonly known in morphometric studies as the *cephalic index* (C.I.) and is calculated according to the following formula.

$$\text{C.I.} = \frac{100\ L_n}{L_t}$$

where L_h is the head length and L_t, the total length. Instead of the total length, the standard length (L_s) can be used.

Fig: 7

153. Hemacanth *

Fr: hémacanthe
Ger:
Lat:
Rus: гемака́нт (pl: гемака́нты)

Sp: hemacanto
See   Hemal spine

154. Hemal arch

Fr: arc hémal
Ger: Hämalbogen
Lat: arcus hæmalis (pl: arcus hæmales); arcus inferioris (pl: arcus inferiores)
Rus: гемáльная дугá (pl: гемáльные дугú)
Sp: arco hemal

The hemal arch is formed during the embryogeny of the fish from the basiventral elements or more rarely from the interventrals of the vertebrae (Fig. 55). In the adult, the branches forming the hemal arch are termed *hemapophyses*. Only the abdominal or caudal vertebrae posses a hemal arch, through which run the dorsal aorta and the caudal vein. The hemal arch is continuous in sturgeons, while in actinopterygians, it is interrupted at each vertebra.

Vertebrae found in archaeological sites are often broken and their arches missing, but even in these cases, it is possible to orient one isolated caudal vertebra, since the hemal arch is broader than its counterpart, the neural arch, and also because both spines are directed backwards. Even when eroded or broken, this arch can be recognized since its bases are further apart than those of the neural arch.

Fig: 35 C

155. Hemal canal

Fr: canal hémal
Ger: Hämalkanal
Lat: canalis hæmalis (pl: canales hæmales)
Rus: гемáльный канáл (pl: гемáльные канáлы)
Sp: canal hemal

A space formed by the branches of the hemal arch in the caudal vertebrae. The dorsal aorta and the caudal vein run through this canal.

Fig: 13 and 35 C

156. Hemal spine

Fr: hémépine
Ger: Hämalstachel; Hämalfortsatz
Lat: processus spinosus ventralis (=inferior) [pl: processus spinosi ventrales
       (= inferiores)]
Rus: гемáльная иглá (pl: г-ные úглы); гемáльний отрóсток
       (pl: г-ные отрóстки)
Sp: espina hemal

The slender bone that joins during the embryogenesis the hemal arch of the caudal vertebrae of teleosts.

Syn: hemacanth; hemapophysis
Figs: 35 C and 37

157. Hemitrich

Fr: hémitriche
Ger: Hemitrich (pl: Hemitrichen)
Lat: hemitrichium (pl: hemitrichia)
Rus:
Sp: hemitrico

     Term applied to each one of the two symmetrical elements that make up a fin ray. The base of each hemitrich enlarges to act as a supporting surface for the fin muscles. These bases embrace the radials of the paired fins, the pterygiophores of the median fins, and the epurals and hypurals of the caudal fin.

158. Heterodont teeth

Fr: dents hétérodontes
Ger: heterodonte Zähne
Lat:
Rus: гетеродо́нтные зу́бы (sing: гетеродо́нтный зуб)
Sp: dientes heterodontos

     Teeth that have different shapes and sizes in the same species, as observed in some sharks (*Heterodontus*) and some teleosts, such as the wolffish (*Anarhichas*). This condition is called *heterodonty*.

159. Holocephalans

Fr: Holocéphales
Ger: Holocephalier
Lat: Holocephali
Rus: слитночерепны́е (=це́льноголо́бные) ры́бы
Sp: Holocéfalos

     Taxonomic group of some 25 species formed with the most advanced cartilaginous fishes (chimaeras and ratfishes), to distinguish them from the sharks and rays. They are edible fishes, but appreciated mainly in economically less developed societies. Their most distinctive skeletal features, from an archaeological point of view, are the presence of
    a) tooth plates instead of individual teeth;
    b) denticles or placoid scales not covering the body, but restricted in their distribution to the head and appendages; and
    c) a poisonous spine on the dorsal fin.

    Tables 2 and 3

160. Holospondylous vertebra   *

Fr: vertèbre holospondyle
Ger: holospondyler Wirbel
Lat: vertebra holospondyla (pl: vertebræ holospondylæ)
Rus: голоспо́ндильный позвоно́к (pl: голоспо́ндильные позвонки́)
Sp: vértebra holospondila

The vertebrae in which all their components (arches, centrum, and spines) are completely fused in a single bone, as in teleost fishes.

161. Holosteans

Fr: Holostéens
Ger: Holosteer
Lat: Holostei
Rus: цельнокостные рыбы; костные ганоиды
Sp: Holósteos

Group established by Müller (1844) to include the Amiiformes and Lepisosteiformes. This term makes reference to the high degree of ossification of their skeleton, although the ossification is not yet complete.

Table   2

162. Homodont teeth

Fr: dents homodontes
Ger: homodonte Zähne
Lat:
Rus: гомодонтные зубы (sing: гомодонтный зуб)
Sp: dientes homodontos

Those teeth of similar shape and size present in the same species. The homodont condition, also called *homodonty,* results from a uniform diet.

Fig: 12

163. Homology

Fr: homologie
Ger: Homologie
Lat:
Rus: гомология
Sp: homología

The concept of homology was introduced in biology when Owen (1848) defined a homologue as "the same organ in different animals under every variety of form and functions" (e.g., the mouth of a crab and the mouth of a fish). This definition does not take into account the concept of evolution that afterwards was to permeate all branches of Biology.

Anatomical structures are the result of genetic and environmental forces acting on each individual and also at the species level, through the process of evolution. Therefore they are dynamic in a double sense: evolutionary or phylogenetic for the group, and embryologic or ontogenetic for the individual. On the one hand, the anatomical features reflect the previous phylogenetic stages and on the other, each individual's features result from ontogenetic stages and behavioral situations.

With this in mind, the concepts derived from anatomical study cannot but reflect this double origin, thereby making the comparison between fossil and extant forms more meaningful. Terms like homology, analogy, convergence, parallelism, primitive or plesiomorphic, specialized, derived or apomorphic, and generalized characters, etc., should be interpreted with evolution in mind. Following this criterion, Simpson (1961) redefined homology as "the resemblance due to inheritance from a common ancestry", to which Smith (1962) added "or ontogenetic development from a common anlage" or primordium. Similar characters involved in the comparison are called *homologous* and the noun applied to them, *homologues* .

Owen (1966) further divided homology into three types:

a) special homology (as it was defined above);

b) serial homology, involving derivation from repetitive anlagen in one organism (e.g., vertebrae); and

c) field homology when the similarity involves one developmental field in two or more species (e.g., the caudal sclerotomes).

164. Horny teeth

Fr: dents cornées
Ger: Hornzähne
Lat:
Rus: роговые зубы (sing: роговой зуб)
Sp: dientes córneos

The horny teeth, derivatives of the epidermis and present in cyclostomes (lampreys), are of a keratinous nature and conical in shape. They are similar in function to the teeth of other vertebrates.

165. Humeral scales *

Fr: écailles humérales
Ger: Humeralschuppen
Lat:
Rus: плечевые чешуи
Sp: escamas humerales

Modified scales with pointed ends found above the pectoral and pelvic fins in many Clupeidae and Engraulidae.

Fig: 6

166. Hyoid arch

Fr: arc hyoïdien
Ger: Zugenbeinbogen; Hyoidbogen
Lat: arcus hyoideus (pl: arcus hyoidei)
Rus: гибидная (=подъязычная) дуга
Sp: arco hioideo

The hyoid arch is the second arch of the viscerocranium, located between the mandibular arch in front and the first branchial arch behind. It consists of several bones of endochondral origin:

a) the uppermost is the *hyomandibular*, with its ventral process which later in evolution becomes independent, and is known as

b) the *symplectic ;*

c) the *interhyal,* a small cylindrical bone that links the upper section of the arch, formed by the hyomandibular and symplectic bones, with its lower section;

d) the *epihyal,* the dorsalmost bone of the lower section of the arch, is joined to the ceratohyal by a suture;

e)   the *ceratohyal,* the largest bone of the lower section of the arch;

f)   in the most ventral and advanced position, one (or more frequently) two hypohyals. In the latter case, the hypohyal placed in a dorsal position is called *dorsohyal,* and the lower one, *ventrohyal ;* and

g)   connected to the ceratohyal and epihyal, a series of variable number of *branchiostegal rays.*

According to the most modern interpretation of this arch, the *ceratohyal* has two centers of ossification during its embryonic development: one dorsal, the *epihyal* (d above) and the other ventral, the *ceratohyal* (e above). The result is that these two bones are considered as one only, but with a double ossification. Consequently, the new nomenclature calls the dorsal ossification *posterior, distal* or *dorsal ceratohyal,* or simply *posterohyal .* The ventral ossification is now named *anterior, proximal* or *ventral ceratohyal,* or simply *anterohyal.*

See   Ceratohyal
Bibl: Nelson 1969
Figs: 1 F, 3, 25 A, and 32

167. Hyomandibular   *

Fr: hyomandibulaire
Ger: Hyomandibulare
Lat: os hyomandibulare (pl: ossa mandibularia)
Rus: гиомандйбула; гиомандибулярная кость
Sp: hiomandibular

Paired endochondral bone corresponding to the dorsal part of the hyoid arch and related to the ventral element of the arch (ceratohyal) by a rod-like bone called *interhyal.* The hyomandibular represents, for some authors, the pharyngobranchial element of the arch, but the interpretation accepted nowadays is that it represents the epihyal. It remains cartilaginous in elasmobranchs and sturgeons.

The hyomandibular is a skeletal element of great phylogenetic importance, owing to its antiquity, since it is found in very primitive fishes, in which it acts as a support for the mandibles. Its shape is generally constant in all groups. Its dorsal part articulates with the otic capsule at the *hyomandibular fossa,* excavated in the adjoining area of the sphenotic, pterotic and prootic. Its ventral part joins the quadrate, reinforced by the symplectic, a bone that was apparently a hyomandibular process, but which became independent in advanced actinopterygians. In fish embryos, the symplectic and the hyomandibular constitute the *hyosymplectic cartilage*

In most teleosts, the hyomandibular has a foramen to allow the passage of the hyomandibular branch of the facial nerve (VII).

Figs: 1 D, 3, 13, 25 A, and 31.
Bibl: Rojo (1986)

168. Hyomandibular foramen

Fr: foramen hyomandibulaire
Ger: Hyomandibular-Loch
Lat: foramen hyomandibulare
Rus: гиомандибулярное отвёрстие
Sp: foramen hiomandibular

The opening through the hyomandibular that permits the passage of the hyomandibular branch of the facial nerve (VII).

Fig: 31

169. Hypobranchials   *

Fr: hypobranchiaux
Ger: Hypobranchialia
Lat: 1. cartilagines hypobranchiales (sing: cartilago hypobranchialis)
       2. ossa hypobranchialia (sing: os hypobranchiale)
Rus: 1. гипобранхиáльные хрящú (sing: г-ный хрящ)
       2. гипобранхиáльные кóсти (sing: г-ная кость)
Sp: hipobranquiales

1. The most ventrally located paired cartilages in chondrichthyans
2. The endochondral bones at the base of the branchial skeleton of osteichthyans. They articulate dorsally with the ceratobranchials and ventrally with the medial series of basibranchials. In teleosts, there are four pairs of hypobranchials since the hypobranchial pair of the fifth arch is always missing. The salmonid fishes have only three pairs of hypobranchials.

Figs: 4, and 25 A

170. Hypohyal   *

Fr: hypohyal (pl: hypohyaux)
Ger: Hypohyale
Lat: os hypohyale (pl: ossa hypohyalia)
Rus: гипохиáльная кость (sing: гипохиáльные кóсти)
Sp: hipohial

Single (*Lepisosteus*) or paired (*Gadus*) bone of endochondral origin, that forms the lower part of the hyoid arch. It articulates dorsally with the ceratohyal and ventrally with the basihyal. Many teleosts have two hypohyals on each side: the dorsal hypohyal or *dorsohyal,* and the ventral hypohyal or *ventrohyal.*

Figs: 51, and 60 A and B.

171. Hypural spine

See Hypurapophysis

172. Hypurals   *

Fr: hypuraux (sing: hypural)
Ger: Hypuralia
Lat: ossa hypuralia (sing: os hypurale)
Rus: гипурáльные кóсточки (sing: г-ная кóсточка) гипурáлии
Sp: hipurales

The hypurals form a series of flat, median bones of subtriangular shape arranged in a fan-like manner. They lie below the caudal section of the vertebral column and articulate with the urostyle (*sensu stricto*) or with the last vertebrae, acting as a support for the caudal fin rays. Their number is large in primitive fishes (*Amia* , 10;

Tarpon, 8; *Albula* and *Salmo*, 7), but in most advanced teleosts, the hypurals fuse into several units, or even one, forming in, this case, the urostyle (*sensu lato*). For a clear understanding of the arrangement, homologies and number of the hypurals, it is necessary to follow their embryonic development in the caudal skeleton.

Gosline (1961) recommended that hypurals be numbered starting with the one in the most ventral position. Previous authors counted them from the top downwards. In this last case, it is easy to misinterpret the homologies between the hypurals of different species, since their tendency to fuse or disappear is more pronounced in the upper ones.

Another problem in the study of homologies is to establish which hypural is the first of the series. Nybelin (1963) has recently determined that the first hypural is the one over which the caudal artery passes immediately after branching from the dorsal aorta. When comparing descriptions of hypurals from different authors, these two criteria should be born in mind.

The bifurcation point of the dorsal aorta marks the boundary between preural and ural vertebrae.

See   Caudal skeleton
Figs: 13 and 37

173. Hypurapophysis

Fr: hypurapophyse
Ger: Hypurapophyse
Lat: hypurapophysis
Rus: гипурапо́физ
Sp: hipurapófisis

This process, already described by Greene and Greene (1914) in chinook salmon (*Oncorhynchus tshawytscha*), has been named and defined by Nursall (1963) as "a lateral process of the anterior hypural bone associated with the *terminal vertebra of Gosline*, serving as the anterolateral portion of the proximal attachment of the hypochordal longitudinal muscle".

The hypurapophysis appears to be a very widespread structure, since it has been found in some 115 species of fish corresponding to 31 families in 10 orders.

Syn: hypural spine (Merriman, 1940)
Fig: 37

174. Illicium *

Fr: illicium
Ger: Köderorgan
Lat: illicium (pl: illicia)
Rus: илли́циум
Sp: ilicio

The illicium is the first dorsal fin ray of some fishes (Lophiiformes, Ceratiiformes) modified into a long appendage ending in a fleshy and sometimes luminous lobe, the *esca* , with which the fish attracts its prey.

The fish is able to extend and retract the illicium at will by the action of protractor and retractor muscles.

175. Incisiform teeth

Fr: dents incisiformes
Ger:
Lat: dentes incisores (sing: dens incisoris)
Rus: резцевйдные зу́бы (sing; резцевйдный зуб)
Sp: dientes incisiformes

Teeth similar to the mammalian incisors and used to cut the prey into pieces. A good example are the razor-sharp teeth of the piranha (*Serrasalmus*). Parrotfishes (Scaridae) have a "beak" which is formed by the incisiform teeth. It is used to cut loose soft particles of coral or to nibble at organisms among the rocks.

Fig: 12 B

176. Incisure

Fr: échancrure
Ger: Einschnitt; Incisur; Inzisur
Lat: incisura (pl: incisuræ)
Rus: вы́резка
Sp: incisura

A natural cleft, notch or fissure on a bone.

Figs: 20 and 23

177. Infraorbitals *

Fr: infraorbitaires
Ger: Infraorbitalia
Lat: ossa infraorbitalia (sing: os infraorbitale)
Rus: нйжнеглазнйчные ко́сти (sing: нйжнеглазнйчная кость)
Sp: infraorbitarios

The infraorbitals form a chain of dermal bones on the inferior border of the eye orbit, starting with the *lacrymal* in front, and ending with the *dermosphenotic* on the posterodorsal region. Included with this name are the so-called *postorbitals*.

The infraorbitals, not to be confused with the suborbitals, vary in number in different fishes, but they are always associated with the sensory infraorbital canal (Stensiö, 1947). This series of bones, found in holosteans and teleosteans, is more constant than the supraorbital series. Generally there are six infraorbitals, although this number is often reduced. The lacrymal is the only one remaining in *Microgadus*.

When counting the infraorbitals, it is recommended to start with the anteriormost (lacrymal or IO_1) and continue in a counter-clockwise direction until the last one, the *dermosphenotic* (IO_6). This is the most accepted way of counting the infraorbitals, although previous authors counted them in a clockwise manner. The second is sometimes called *jugal*.

In some works, these bones are numbered as $SO_1 -- SO_6$ because they were considered to be suborbitals.

See Circumorbitals, Suborbitals and Supraorbitals
Figs: 1 A and 14 A

178. Infrapharyngobranchial dental plate

Fr: plaque dentaire infrapharyngobranchiale
Ger: infrapharyngobranchiale Zahnplatte
Lat:
Rus: инфрафарингобранхиа́льная зубна́я пласти́на
Sp: placa dental infrafaríngea

The upper surface of the fifth ceratobranchial is covered by a dentigerous plate of dermal origin called the *infrapharyngobranchial dental plate*. Sometimes the plate is divided into two sections with teeth of two different types, as in *Merluccius* (Rojo, 1976). The assemblage of the fifth ceratobranchial, the plate and the teeth is called simply *fifth ceratobranchial*.

See   Pharyngobranchials
Syn: lower pharyngeal bone
Fig: 4

179. Innominate   bone

Fr: os inconnu
Ger: unbenannter Knochen
Lat: os innominatum (pl: ossa innominata); os coxae (pl: ossa coxæ)
Rus: безымя́нная кость (pl: безымя́нные ко́сти)
Sp: hueso innominado

DeKay (1842) gave this name to the bone of the pelvic fins in sticklebacks.

Syn: posterior process (Nelson, 1971); pubic bones (Gunther, 1859); pelvic plate (Regan, 1909); pelvic bone (Bertin, 1944); medial plate (Pencsak, 1965)

180. Intercalar   *

Fr: intercalaire
Ger: Intercalare
Lat: os intercalare
Rus: интеркаля́рная кость
Sp: intercalar

Name proposed by Vrolik (1873) for the tendinous ossification that forms the posterior wall of the otic capsule, between the prootic, the pterotic, and the exoccipital. Its position is variable, but it is always associated with the ventral process of the posttemporal.

The intercalar either extends forward to meet the prootic (plesiomorphic condition) or remains disconnected (apomorphic condition). Its loss represents a more specialized condition, related to the reduction of the bones of the pectoral girdle or fin.

Although this bone is still called opisthotic by many authors, it is recommended that this name be replaced by the more appropriate one *intercalar*. This bone is missing in Siluridae (Lundberg, 1975). In Gadidae, the intercalar has a foramen for the glossopharyngeal nerve (IX).

Fig: 2

181. Interhyal   *

Fr: interhyal
Ger: Interhyale
Lat: os interhyale (pl: ossa interhyalia)
Rus: интерхиáльная кость (pl: интерхиáльные кóсти)
Sp: interhial

Small bone of cylindrical shape, that joins the hyomandibular with the posterohyal (=epihyal). This named was proposed by Norman (1926) when he demostrated that it is not homologous with the *stylohyal* of tetrapods. Nevertheless, many authors still use this last name. The interhyal is an endochondral paired bone characteristic of bony fishes. Its evolutionary origin has not yet been clarified: Stöhr (1882) considers it to be a segment of the ceratohyal; Allis (1922) thinks that it is the epihyal or dorsal part of the hyoid arch; while Holmgren and Runnström (1943) suspect a double origin, partly from the primary hyoid arch and partly from an epihyal branchial ray. Lately, Véran (1988) considered this bone as the independent ossification of the lower part of the dorsal branch of the hyoid arch.

The interhyal bone is found in all actinopterygians, but is missing in dipnoans, in which the palate is fixed.

Syn: stylohyal (Sewerstzoff, 1928)
Figs: 1 F, 3, 25 A, and 32

182. Intermuscular   bones

Fr: os intermusculaires
Ger: intermuskuläre Knochen
Lat: ossa intermuscularia (sing: os intermuculare)
Rus: межмы́шечные кóсти (sing: межмы́шечная кость)
Sp: huesos intermusculares

See   Metaxymiosts

183. Interopercle   *

Fr: interoperculaire
Ger: Interoperculum
Lat: os interoperculum (pl: ossa interopercula)
       os interoperculare (pl: ossa interopercularia)
Rus: межкры́шечная (=интероперкуля́рная) кость
       [pl: межкры́шечные (= интероперкуля́рные) кóсти]
Sp: interopérculo; interopercular

Laminar dermal bone that occupies the ventral part of the opercular membrane. It is joined to the lower mandible by a ligament, as in many fossil fishes (Acanthodii). It is missing in *Lepisosteus* and in some teleosts, such as Siluridae.

There are several hypotheses regarding its origin. Schaeffer and Rosen (1961) consider the interopercle as a modified branchiostegal ray, while Regan (1929) is of the opinion that it derives from the subopercle of fossil actinopterygians. Finally, Casier (1954) sees it as the modification of the superior gular plate.

Syn: interopercular
Figs: 1 C, 3, 13, 14 B, and 24

184. Ischiopubic bar

Fr: cartilage ischio-pubien; barre ischio-pubienne
Ger:
Lat:
Rus: седа́лищнолобко́вый хрящ
Sp: barra isquiopúbica

A straight cartilaginous bar that constitutes the pelvic girdle in cartilaginous fishes. It extends horizontally, joining both pelvic fins which articulate at the acetabular surfaces. An iliac process extends from each end of the bar.

In all living elasmobranchs, the right and left halves of the pelvic girdle are fused into a single ventral puboischiadic bar (Compagno, 1973).

Syn: ischiopubic cartilage; puboisquiadic bar

185. Isospondylous fishes

Fr:
Ger: isospondyle Fische
Lat: Isospondyli
Rus:
Sp: peces isospondilos

Under the name Isospondyli, Regan (1909) grouped the fishes having the vertebrae immediately after the skull similar in shape to the remaining ones, in contrast to the ostariophysans, in which the anterior vertebrae are greatly modified. Modern classifications have rejected this artificially constructed group, and the fishes previously assigned to it have been distributed among different orders (Clupeiformes, Osteoglossiformes, Salmoniformes, Cetomimiformes, etc.).

Bibl: Greenwood *et al.* (1966)

186. Jugal

Fr: jugal
Ger: Jugale; Jochbein
Lat: os jugale
Rus: скулова́я кость (pl: скуловы́е ко́сти)
Sp:–   yugal

1. Name given by some authors to the supramaxilla.
2. This name is also sometimes applied to the second infraorbital bone, the lacrymal being the first.
3. In early fish literature, this name was applied to the lacrymal.

187. Jugostegal rays

Fr: rayons jugostégaux (sing: rayon jugostégal)
Ger: Jugostegalstrahlen
Lat: ossa jugostegalia (sing: os jugostegale)
Rus:
Sp: radios yugóstegos

Name given by Parr (1930) to the very slender ossifications that extend from the upper lateral part of the gill cover on each side to its lower part in the anguillid genus *Myrophis*. They are located between the branchiostegal rays and the gill openings and are present in large numbers (36-46). They are also present in Echelidae, Ophichthyidae, and Neenchelyidae families.

188. Jugular fishes

Fr: poissons jugulaires
Ger:
Lat: pisces jugulares
Rus:
Sp: peces yugulares

Taxonomic group created by Linnaeus (1758) to assemble those bony fishes which have the pelvic fins in front of the pectoral fins, as in the cod family (Gadidae).
Table   3

189. Kinethmoid   *

Fr: kinethmoïde
Ger: Kinethmoid
Lat:
Rus:
Sp: kinetmoides

Name proposed by Harrington (1955) for the median bone located on the antero-dorsal part of the ethmoidal region of many teleosteans. This term should replace the term "rostral" which is now mostly used for a bone present in fossil and primitive bony fishes. The term " kinethmoid" or *moveable ethmoid* , refers to its involvement in the protrusion mechanism of the premaxillaries.

190. Lacrymal *

Fr: lacrymal
Ger: Tränenbein; Lakrymale
Lat: os lacrimale (pl: ossa lacrimalia)
Rus: слёзная (=лакримáльная) кость [pl:с-ные (=л-ные) кóсти]
Sp: lacrimal

The anteriormost bone of the infraorbital series.that encloses the cephalic section of the infraorbital canal. It is a paired, dermal bone of larger size than the remaining ones of the series

Syn   lacrimal; lachrymal; praeorbital (Berg, 1940)
Figs: 1 A, 13, 14 A, and 21

191. Lacrymojugal

Fr: lacrymo-jugal
Ger: Lakrymojugale
Lat: os lacrimojugale (pl: ossa lacrimojugalia)
Rus:
Sp: lacrimoyugal

Paired dermal bone resulting from the fusion of the first two infraorbitals in *Latimeria*. This bone encloses the anterior part of the infraorbital canal. Ramaswami (1948) also found it in the Homalopteridae.

192. Lagena

Fr:	lagena
Ger:	Lagena
Lat:	lagena (pl: lagenæ)
Rus:	
Sp:	lagena

The chamber of the membraneous labyrinth located in the lowest and outermost position. It forms part of the saccule or is a simple outgrowth of it and it is looked upon as a rudimentary *cochlea*. It is not well developed in fishes, except in Cypriniformes.

Inside the lagena of fin-rayed fishes, there is a concretion of calcium carbonate (the earstone or otolith) known as *lagenolith* or *asteriscus,* so-named because of its star-like shape.

It is believed that the lagenar macula is responsible for the sense of hearing in fishes, since the lagena becomes the auditory organ after increasing in size in some vertebrates (reptiles and birds) and coiling to form the cochlear duct in mammals.

See	Asteriscus
Fig:	42

193. Lapillus

Fr:	lapillus
Ger:	Lapillus; Utrikulolith
Lat:	lapillus (pl: lapilli)
Rus:	
Sp:	lapilo

The lapillus is the otolith located in the utriculus of the actinopterygian fishes. It is held in a vertical position and leans against the cells of the utricular macula, which are innervated by the anterior branch of the eighth cranial nerve (stato-acoustic nerve). The movements of the lapillus inform the brain of the position of the fish in the water and help it to maintain its equilibrium. It is very small in most fishes, except in the Siluridae, in which it is the largest of the three otoliths.

Syn:	utriculith
Figs:	42 and 45

194. Lateral ethmoid   *

Fr:	ethmoïde lateral
Ger:	Ethmoideum laterale
Lat:	os ethmoideum laterale (pl: ossa ethmoidea lateralia)
Rus:	боковая обонятельная (= решётчатая) кость
Sp:	etmoides lateral

Paired bone of perichondral origin, that arises from the *lamina orbitonasalis*, according to de Beer (1937), and which is present in *Amia, Acipenser*, and teleosts.

Syn: pleurethmoid; prefrontal; parethmoid; exethmoid; ectecthmoid (name to be rejected since according to de Beer (1937), this bone is only present in birds).
See   Prefrontal
Figs: 2, 13, and 14 A

195. Lateral gular

Fr: plaque gulaire latérale
Ger: Gularia lateralia
Lat: ossa gularia lateralia (sing: os gulare laterale)
Rus: боковáя гулярная пластúнка (pl: б-ы́е г-ные пластúнки)
Sp: placa gular lateral

Paired dermal bone located at the base of the throat of some primitive fishes (*Latimeria*)

See Gular plate

196. Lateral line

Fr: ligne latérale
Ger: Seitenlinie
Lat: linea lateralis (pl: lineæ laterales)
Rus: боковáя лúния
Sp: línea lateral

In the strictest sense, the lateral line consists of the canals or grooves that run along the side of the fish from the opercular membrane, where it connects with the head canals, down to the caudal peduncle. In most cases, the lateral line is an uninterrupted straight line, but very often its anterior part makes a deep arch (*Selene vomer*), and occasionally, it is cut into sections, as in the radiated shanny (*Ulvaria subbifurcata*).

The scales covering the canal have in the center an opening, sometimes prolonged into a tube or a funnel running obliquely and through which water can enter the underlying canal. The neuromasts of these canals can detect changes in water pressure, giving the fish information about its depth, the proximity of prey, the presence of predators and stationary objects, and the position of other fishes in the school. The neuromasts of the lateral line are innervated by the sensory neurons of the lateral branch of the vagus nerve (X).

The position, shape, and angle of the tubes of these scales as well as the number of scales, can be a specific characteristic used in the identification of fish remains.

Fig: 6

197. Lateral line scales

Fr: écailles de la ligne latérale
Ger: Seitenlinienschuppen
Lat: squamæ lineæ lateralis
Rus: чешýи боковóй лúнии
Sp: escamas de la línea lateral

The scales related to the lateral line system are characterized by the presence of a central pore, which in some cases extends into a cylindrical or funnel like canal. Water enters through this opening to the sensory canal system that runs under the skin.

Fig: 6

198. Lepidotrichs

Fr: lépidotriches
Ger: Lepidotrichen (sing: Lepidotrich)
Lat: lepidotrichia (sing: lepidotrichum)
Rus: лепидотри́хии (sing: лепидотри́хия)
Sp: lepidotricos

Rays found in the median and paired fins of the actinopterygian fishes. Their origin is exclusively dermal. They are made of bundles of collagen fibers surrounded by a bony matrix, exuded by cells very similar to osteoblasts.

They develop in the embryo after the actinotrichs are formed, which they displace and finally replace in some cases. They are composed of two pieces (*hemitrichs*) joined along their whole length except at their bases, where they open to embrace the pterygiophores.

Goodrich (1904) concluded that they were derived from scales, and called them *lepidotrichia*. A lepidotrich when fully developed can be simple, segmented or ramified.

Fig: 13

199. Leptoid scales

Fr: écailles leptoïdes
Ger: leptoide Schuppen
Lat:
Rus:
Sp: escamas leptoideas

See   Scales
Table   4

200. Lingual plate

Fr: plaque linguale
Ger: Zungenplatte
Lat: os linguale plattum
Rus: языкова́я (= язычко́вая) пласти́нка
Sp: placa lingual

The dermal toothed bone that covers and sometimes fuses with the underlying basihyal. The Osteoglossidae are characterized by the presence of a strong lingual plate bearing teeth, as the family name implies.

Syn: glossohyal; dermentoglossum (de Beer, 1937); supralingual (Tchernavin, 1938); basihyal dental plate.

Fig: 34 A

201. Lower pharyngeal   *

Fr: pharyngien inférieur
Ger: unterer Schlundknochen
Lat: os pharyngeum inferior (pl: ossa pharyngea inferiora)
Rus:
Sp: faríngeo inferior

Name applied to the fifth ceratobranchial, especially in cyprinoid fishes, in which it attains a large size and usually bears strong teeth.

See   Ceratobranchials and Infrapharyngobranchial plate
Fig: 33

202. Malacopterygians

Fr: Malacoptérygiens
Ger: Malacopterygier; Weichflosser
Lat: Malacopterygii
Rus: мягкопёрые рыбы
Sp: Malacopterigios

Group of actinopterygian fishes, which, as their name implies, is characterized by having only flexible, soft fin rays, as in Salmonidae and Cypriniformes. This feature represents a more primitive character than that of acanthopterygians, which have in addition some spiny rays.   This taxonomic division was established by Willughby (1686) in his *Historia piscium.*

203. Mandibular arch

Fr: arc mandibulaire
Ger: Mandibularbogen
Lat: arcus mandibularis
Rus: челюстная дугá (pl: челюстные дугú)
Sp: arco mandibular

In a strict sense, the term *mandibular arch* is applied to an assemblage of cartilages and bones (the palatoquadrate cartilages in the upper jaw and Meckel's cartilages, the angular and other less important bones, in the lower) all belonging to the visceral skeleton. This assemblage forms the core of fish jaws and it is also referred to as *primary mandibles* .

Associated with this arch are teeth and dermal bones which are not considered part of the mandibular arch, since they joined the mandibles much later in the evolution of fishes. Nevertheless, these later elements also form part of the mandibles, and for this reason they are called *secondary mandibles* .

The name "arch" given to the mandibles arises from the most accepted evolutionary interpretation of its formation, i. e. , that the mandibles of the primitive jawed fishes (placoderms) derive from the first -- or one of the first -- branchial arches of jawless fishes (ostracoderms).

In cartilaginous fishes, the mandibular arch forms a complete arch, composed of two upper bars --*palatoquadrate* or *pterygoquadrate cartilages* -- which join at the maxillar symphysis, and form the upper mandible, also called *maxilla* , and two lower bars (*Meckel's cartilages*) fused at the mandibular symphysis, thereby forming the lower jaw or *mandible* proper.

In bony fishes, these cartilages developed several centers of ossification, giving origin to endochondral bones (autopalatine, quadrate, metapterygoid, articular, etc.). To these endochondral bones, bones of dermal origin associated with teeth, among them the premaxilla, maxilla, dentary, dermopalatine, etc. were later added during the evolution of this group of fishes.

Bibl: Berry (1964); Dobben (1935); Eaton (1935); Schaeffer and Rosen (1961)

204. Mandibular symphysis

Fr: symphyse mandibulaire
Ger: Unterkiefer-Symphyse
Lat: symphysis mandibularis
Rus: мандибулярный симфиз; мандибулярное сращенйе
Sp: sínfisis mandibular

The plane of contact between the right and left mandibles.

205. Maxilla   *

Fr: maxillaire
Ger: Oberkieferbein
Lat: os maxillare (pl: ossa maxillaria); maxilla (pl: maxillæ)
Rus: челюстнāя (=максиллярная) кость [pl: ч-ны̄е (=м-ные) к.]
Sp: maxila; maxilar

Paired dermal bone of the upper jaw located posterior to the premaxilla. Its evolution presents the following trends:
1)   it is a long bone in fossil bony fishes, its length diminishing in modern ones;
2)   it is a toothed bone in the fossil forms retaining the teeth in the most primitive modern teleosts (Elopiformes, Anguilliformes, Clupeiformes and Osteoglossomorpha), with a tendency towards losing them in more advanced teleosts (from Salmoniformes to Pleuronectiformes);
3)   in the early evolutionary stages, the maxilla plays a very important role in the capture of food since it forms the mouth gape, a condition that is retained in primitive modern teleosts. In the remaining modern teleosts, there is a tendency for the maxilla to be pushed backward because of the growth of the premaxilla, until it is finally excluded from the mouth gape; and
4)   in fossil osteichthyans the maxilla is closely related to the dermal bones of the palatoquadrate bar and is therefore without movement, but in *Amia,* it is already independent and moves freely, characteristic which is maintained in the remaining actinopterygians.

Syn: maxillary
Figs: 1 B, 3, 13, 14 B, and 16
Bibl: Dobben (1935); Rojo (1986)

206. Maxillo-infraorbital

Fr: maxillo-infraorbitaire
Ger: Maxilloinfraorbitale
Lat:
Rus:
Sp: maxilo-infraorbitario

According to Pehrson, the maxillo-infraorbital is a paired bone formed by the fusion of two ossifications: one in the lower part of the orbit, corresponding to the jugal, and the other below this, close to the edge of the mouth.

207. Meckel's cartilage

Fr: cartilage de Meckel
Ger: Meckelscher Knorpel
Lat: cartilago meckeli (pl: cartilagines meckeli)
Rus: мёккелев хрящ
Sp: cartílago de Meckel

Cartilage, described by Meckel (1820), that forms the lower mandible of fish embryos. This cartilage is also known as *mandibular cartilage* or *primary mandible.*

Meckel's cartilage is absent in lampreys and hagfishes, which lack mandibles altogether. It appears in its full development and function with the first gnathostomes (Placodermi), acting as support and attachment for the mandibular muscles and teeth. Both cartilages meet in the mandibular symphysis. Together with the palatoquadrate cartilage, they form the mandibular arch.

The tendency to trace homologies between the mandibular and branchial arches has resulted in Meckel's cartilage being compared to the pharyngobranchial, the epibranchial, the ceratobranchia,l and the hypobranchial. The most accepted interpretation is that it can be homologized clearly with the complex cerato-hypobranchial.

Meckel's cartilage attains its maximum development in Chondrichthyes, where it remains cartilaginous during the whole life of the fish. In bony fishes, Meckel's cartilage is a cylindrical rod of cartilage which extends directly forward from a process located at the angle between the dorsal and ventral branches of the angular, and fits into the space between the two arms of the V-shaped dentary. In chondrosteans (sturgeon and paddlefishes) it also remains cartilaginous for the life of the fish. A special situation arises in *Amia*, where Meckel's cartilage ossifies into several centers, which are difficult to homologize with those in teleosts. On the posterior part of Meckel's cartilage, four ossicles are formed (*a, b, c,* and *d*), known also as *Bridge's ossicles* . These little bones could represent the remains of a large bone, present in fossil paleoniscoids. The most accepted homology between these ossicles and those found in Teleosts is as follows: ossicle *a* represents the retroarticular; *b* and *c* combined are equivalent to the articular, and *d* would represent the coronomeckelian.

In some teleost fishes, Meckel's cartilage forms several centers of ossification: mentomeckelian, mediomeckelian, coronomeckelian, articular and the retroarticular in some teleosts. In most fishes, however, it remains only as a pointed rod sheathed by the dentary and angular dermal bones.

Syn: mandibular cartilage; ceratomandibular cartilage
Fig: 18

208. Median fins

Fr: nageoires impaires
Ger: unpaarige Flossen
Lat: pinnæ medianæ (=impariles; impares); perissopterygium (pl: perissopterygia)
Rus: срединные плавники
Sp: aletas impares

The median fins (*perissopterygia* [sing: *perissopterygium*]) are located in a vertical position along the middle plane of the body. Their names refer to the relative position they occupy: the dorsal fins, on the back between the nape and the tail; the caudal fin, forming the tail; and the anal fin, placed on the ventral side immediately behind the vent. The main function of the median fins is to stabilize the fish as it swim, preventing rolling movements. In the evolution of fishes, the median fins preceded the paired fins, but disappeared without any trace in the tetrapods.

Figs: 6 and 13
Bibl: Eaton (1945); Goodrich (1906)

209. Membrane bones

Fr: os de membrane
Ger: Hautknochen; Deckknochen
Lat: ossa investientia (sing: os investientium)
Rus: покрóвные кóсти (sing: покрóвная кость)
Sp: huesos de membrana

Reichert in 1838 was the first to recognize the character of membrane bones, as those formed by direct ossification of the connective tissue layers of the mesenchyme of the dermis. The result is always a thin, laminar bone located near the surface of the body. Membrane bones are found in the head, mandibles, buccal cavity and in the pectoral girdle of the fish. The scales represent the last stages of the dermal bones of fossil ostracoderms and placoderms.

The osteogenesis of membrane bones is a metaplasic process, since it is due to the gradual transformation of connective mesenchymous tissue into bone. These bones do not differ histologically from the bones formed from a previous cartilage template.

Although the terms *membrane bones* and *dermal bones* are not strictly synonymous, they are interchangeable, especially in reference to skull bones and those of the pectoral girdle.

See   Dermal bones
Syn: achondral bones; investing bones
Table   1

210. Membraneous labyrinth

Fr: labyrinthe membraneux
Ger: häutiges Labyrinth
Lat: labyrinthus membranaceus (pl: labyrinthi membranacei)
Rus: перепóнчатый лабирúнт
Sp: laberinto membranoso

Term applied to the assemblage of the semicircular canals and the otosac, since it reflects the membraneous nature of their walls. The canals and the chambers (utricle, saccule and lagena) are filled with *endolymph* and are bathed on the outside by the *perilymph*. In cyclostomes and cartilaginous fishes, the membraneous labyrinth is enclosed in a cartilaginous capsule, while in bony fishes it is completely encased by several bones which form the so-called *bony labyrinth* .

According to Lowestein (1957), the labyrinth maintains and regulates muscle tone, and acts as a receptor for angular acceleration, gravity, and sound.

Fig: 42
Bibl: Gray (1951)

211. Meningost

Fr: méningoste
Ger:
Lat
Rus:
Sp: meningosto

Name proposed by Chabanaud (1936) for the lateral ossification of the basisphenoid.

See Basisphenoid

212. Mentomeckelian   *

Fr: mentomeckélien
Ger:
Lat: os mentomeckelium (pl: ossa mentomeckelia)
Rus: подбородочночёлюстная (= подбородочная) кость
Sp: mentomeckeliano

Small paired bone of endochondral origin formed on the anterior end of the Meckel's cartilage. It is found in *Amia,* and according to some authors (de Beer, 1937), is also found in the larval stages of salmon. In most teleosts it fuses early with the dentary, except in adult Cyprinidae, in which it remains independent.

Syn: mentomandibular ossification (Holmgren and Stensiö, 1936; Lekander, 1949)
Fig: 25 A

213. Meristic characters

Fr: charactéristiques méristiques
Ger: meristische Merkmale
Lat:
Rus: мeристи́ческие (=счётные) при́знаки
Sp: caracteres merísticos

Characters or characteristics are those features repeated serially in fishes (fin rays, vertebrae, scales, etc.) and those organs and their products which appear in a certain quantity, and are consequently capable of being counted (pyloric caeca, etc.).

The most important meristic characters used to identify taxonomic groups, such as species, populations, and races, are: soft and spiny fin rays, spines, finlets, scales, scutes, branchiostegal rays, buccal and pharyngeal teeth, vertebrae, ribs, pyloric caeca, myomeres, and photophores.

Since all these features show a degree of variability within a taxonomic group, it is necessary to give the minimum and maximum number found, the mean, mode, median, the standard deviation, the coefficient of variability, all with their respective standard errors and fiducial limits. In order to assess the dependability of these values, it is necessary to know the size of the sample from which they are derived.

The data obtained can be compared by statistical tests of significance with similar data from other populations, with the objective of establishing the degree of affinity between them. Before comparing two groups, one must determine whether the meristic character selected varies with the size or the age of the fish. If it does, the populations compared should be composed of fishes of similar size and/or age.

The minimum, maximum, and modal values of a meristic character for each species depend on a very complex set of genetic and environmental factors. As a result, it is necessary to establish experimentally what degree of variability is attributable to each factor.

The environmental factors, whose influence on the meristic characteristics is best established during the embryonic stages, are: water temperature, salinity, intensity of light, geographical altitude and latitude, oxygen and carbon dioxide concentration, and pH. Although their influence is well known in many cases, the mechanisms whereby they operate remain to be discovered. It should also be noted that the influence of these environmental factors varies not only according to their intensity, but also to their duration and frequency.

Owing to the ecophenotypical plasticity of the meristic characters, some doubts have been raised about the value of the results obtained from observational studies without the support of controlled experiments, or without observing their geographic variation under different climatic regimes.

The genetic component is also suspect, since the progenitors of a population vary widely among themselves. The results obtained from a sample represent only the contribution of one group of parents, and therefore, are less and less representative for a species as the sample gets smaller. The smaller the sample, the higher the sampling error, which is compounded by the possible influence of the genetic drift.

Bibl: Hubbs (1922); Lindsey (1955); Lindsey and Ali (1965); Rojo and Capezzani (1971)

214. Mesethmoid

Fr: mésethmoïde
Ger: Mesethmoid
Lat: os mesethmoideum (pl: ossa mesethmoidea)
Rus: межобонятельная (= средняя обонятельная) кость; мезэтмоид
Sp: mesetmoides

Name that should be rejected from fish anatomy, since the bone so named is not homologous to the mesethmoid of the mammalian skull (Harrington, 1955). The fish bone should be called *ethmoid*.

See   Supraethmoid

215. Mesocoracoid   *

Fr: mésocoracoïde
Ger: Mesocoracoid
Lat: os mesocoracoideum (pl: ossa mesocoracoidea)
Rus: мезокоракоид
Sp: mesocoracoides

Paired endochondral bone of the pectoral girdle, located between the scapula and the coracoid, present in Clupeiformes and Cypriniformes, among other orders of fishes.

Fig: 38 C

216. Mesonost *

Fr: mésonoste
Ger:
Lat:
Rus:
Sp: mesonosto

The middle element of the pterygiophores. This little bone is missing in many teleosts.

See   Pterygiophores
Fig: 41 B

217. Mesopterygium

Fr: mésoptérygium
Ger: Mesopterygium
Lat: mesopterygium (pl: mesopterygia)
Rus: мезоптеригий; срединный базальный хрящ
Sp: mesopterigio

The cartilage that occupies the middle position of the three basal cartilages in the pectoral fin of Chondrichthyes and primitive Teleostomi. The mesopterygium is absent in the pelvic fins.

218. Mesopterygoid

Fr: mésoptérygoïde
Ger: Mesopterygoid
Lat: os mesopterygoideum (pl: ossa mesopterygoidea)
Rus: средняя крыловидная кость; мезоптеригбид
Sp: mesopterigoides

See   Endopterygoid

219. Metapterygium

Fr: métaptérygium —
Ger: Metapterygium
Lat: metapterygium (pl: metapterygia)
Rus: метаптеригий; дистальный базальный хрящ
Sp: metapterigio

The innermost basal cartilage of the paired fins in Elasmobranchii and Holocephali. It is usually the longest of the three basal cartilages. The metapterygium is exceptionally long in the pelvic fins, especially in the males, in which its distal part is modified into a copulatory organ, called *clasper*.

220. Metapterygoid *

Fr: métaptérygoïde
Ger: Metapterygoid

Lat: os metapterygoideum (pl: ossa metapterygoidea)
Rus: метаптеригбид
Sp: metapterigoides

Paired bone of endochondral origin, formed in the posterior part of the palatoquadrate cartilage and corresponding, accordingly, to the primary mandible. The metapterygoid is located in the angle formed by the two branches of the suspensorium. It is lost in many groups of modern teleosts.

Syn: mesopterygoid
See   Mandibular arch
Figs: 1 D, 3, 13, 25 A, and 30

221. Metaxymiosts   *

Fr: métaxymiostes
Ger: Metaxymiostier
Lat:
Rus:
Sp: metaximiostos

Name proposed by Chabanaud for the membrane bones formed in the intermuscular septa. Since they are not related to the vertebrae, they should not be confused with true ribs. They are very well developed in Clupeidae, occupying three positions: the dorsal or *epineural bones* related to the neural spines, the central or *epicentral bones*, and the ventral or *epipleural bones*, related to the ventral ribs.

Syn   intermuscular bones

222. Molariform teeth

Fr: dents molariformes
Ger: molariforme Zähne
Lat: dentes molariformes (sing: dens molariformis)
Rus: коренновидные зубы (sing: коренновидный зуб)
Sp: dientes molariformes

Fishes feeding on molluscs and crustacea usually have teeth similar to the mammalian molars. The crown is round in outline and flattened, an appropriate shape for crushing and grinding shells. These teeth are found in many bottom-dwelling fishes (*Raja, Anarhichas,* Sparidae, and Sciaenidae.

223. Monospondylous vertebra   *

Fr: vertèbre monospondyle
Ger: monospondyler Wirbel
Lat: vertebra monospondyla (pl: vertebræ monospondylæ)
Rus: моноспондиьнáьный позвонóк (pl: м-ные позвонкй)
Sp: vértebra monospondila

Vertebra with only one centrum, corresponding embryologically to one sclerotome. This type of vertebra is the normal type in Chondrichthyes and Actinopterygii.
Fig: 35

224. Morphometric characters

Fr: charactéristiques morphométriques
Ger: morphometrische Merkmale
Lat:
Rus: морфометрйческие прйзнаки (sing: м-кий прйзнак)
Sp: características morfométricas

    Morphometric characters are those which can be measured and expressed by a numerical value. These characters can be linear (total length, head length, eye diameter, etc.), bidimensional (the absorption surface of the digestive tract, etc.), tridimensional (volume, etc.), or a derivative of the previous ones (weight, density, swimming speed, etc.).

    Their value is important in biometric studies, since they provide a means to assess the biological variables through the vital cycle of the fish and also to estimate the degree of variability for a character within a population, race or species. Therefore, it is possible to characterize a taxonomic group statistically by measuring a fish sample and estimating its parameters (range, index of variability, mean, median, mode, standard deviation, variance, errors, etc.).

    Besides the individual or populational value of the data obtained, the morphometric characters, through their numerical values, allow meaningful comparison between two or more individuals, sexes, populations, or different life stages of the same individual. These comparisons permit the establishment of mathematical relationships between the groups, by means of the calculation of the degree of interdependence among them by the correlation coefficient, and the nature of this same relationship by the equation of regression.

    Because many morphometric characters (weight, length, volume, number of eggs, etc.) are related to size, and because this is, in turn, a function of time and food, the values of the morphometric characters are used to estimate the absolute and relative growth either of the fish or any of its organs.

    In archaeological research, the most useful characters that can be related to the live size of the fish are bone, tooth, scale, and otolith dimensions. In order to obtain the live size of the fish, regression equations must be prepared from a reasonable number of bones of the same species. Any bone measurement can be related to fish length by a regression equation of the linear type

$$Y \text{ (fish length)} = a \text{ (constant)} + b\,X \text{ (bone measurement)}.$$

    When calculating the weight, it should be kept in mind that the relationship between the fish weight and any linear dimension of a bone, is exponential and it is represented by the following equation

$$W \text{ (fish weight)} = a\,L^n$$

where a is a constant, L the linear dimension, and n an exponent whose value varies from 2 to 4, but in most fishes it oscilates around the value 3. (See Weight). To simplify the weight calculation, both the weight and the linear measurement should be converted into their corresponding logarithms. The regression line using these new values is once again linear.

    In all cases, the correlation coefficient between each pair of measurements (fish length -- bone dimension, and fish weight -- bone measurement) should be calculated to see whether a strong and consequently, valid relationship, exist or not. The closer the value of r (correlation coefficient) is to unity, the more valuable is the regression obtained for estimating the fish live size. The lower the value of the correlation coefficient, the larger the error in the calculation values.

Fig: 7
Bibl: Rojo and Capezzani (1971)

225. Myodome   *

Fr: myodome
Ger: Myodom
Lat:
Rus: миодо́м
Sp: miodomo

Weitzman (1967) gave this name of the median bone of perichondral origin, formed in the ethmoid cartilage. The myodome, a small, thin bone of subconical shape, forms the anterior wall of the myodome cavity.

Fig: 14 A

226. Myodomes

Fr: myodomes
Ger: Myodome
Lat:
Rus: миодо́мы
Sp: miodomos

Cavities formed in the fish's skull to provide a point of insertion for the motor muscles of the eye. Fishes have two myodomes: the anterior, excavated in the ethmoid region, from which the oblique muscles (superior and inferior) arise, and the posterior myodome, which anchors the four straight muscles of the eye (anterior, posterior, superior and inferior). This last part is formed by a space limited above by the floor of the oto-occipital cavity and below, by the basisphenoid or the parasphenoid.
    The anterior myodome is sometimes divided into two cavities, dorsal and ventral. The posterior myodome divides, in its turn, into left and right cavities on the sides of the hypophyseal fossa. The ventral base of the basisphenoid separates those two cavities.
    The actinopterygians have well developed myodomes, although in some groups, such as Siluridae, *Lepisosteus*, and *Acipenser*, they have been lost or probably were never present.

227. Myxopterygium

Fr: myxoptérygie
Ger: Kopulationorgan
Lat: myxopterygium (pl: myxopterygia)
Rus: миксопте́ригий
Sp: mixopterigio

In males of both viviparous and oviparous cartilaginous fishes, the pelvic fins are modified into an intromittent organ to assure fertilization of the ova, which always occur inside the oviduct. Supplemental claspers are present in chimaeroids: one in front of each pelvic fin (prepelvic tenaculum) and another on the forehead (cephalic tenaculum).

The skeleton of the pelvic myxopterygium or clasper consists essentially of the following cartilages: the long stem cartilage; the shorter distal cartilage; the dorsal terminal cartilage or rhipidion, which is curved and pointed; the hook-shaped dorsal terminal cartilage or claw with a cutting edge on its inner margin; and the narrow and sharp spur. These last two cartilages move in opposite directions and anchor the clasper inside the female oviducts.

Besides these major elements there are several small cartilages, variable in number, shape, and position, that join the clasper with the pelvic fin proper.

A groove running along the clasper guides the sperm into the oviduct. At the same time, sea water squeezed from the siphon sac dilutes the sperm and facilitates its transport.

Syn: clasper, tenaculum

228. Nasal   *

Fr: nasal
Ger: Nasale
Lat: os nasale (pl: ossa nasalia)
Rus: носовáя кость (sing: носовы́е кóсти)
Sp: nasal

Paired dermal bone developed between the two nares, through which part of the supraorbital canal passes. The nasal bone is located in a dorsal position on the ethmoid region of the fish head. It meets the antorbital posteriorly, or in its absence, the lacrymal. Some fishes, such as *Lepisosteus*, have three nasals, the nomenclature of which is not uniform. These three nasals are commonly named (from front to back), *prenasal, adnasal,* and *premaxillo-nasal,* while Aumonier (1942) calls them *rostral, nasal,* and *antorbital,* respectively.

Syn: naso-postrostral (Holmgren and Stensiö, 1936)
Figs: 1 A, 13, 14 B, and 21

229. Nasal process

Fr: processus nasal
Ger: Nasalfortsatz
Lat: processus nasalis (pl: processus nasales)
Rus: носовóй отрóсток
Sp: proceso nasal

Patterson (1973) proposed the application of this term to the vertical expansion of the premaxillae in holosteans.

230. Naso-premaxilla

Fr: naso-prémaxillaire
Ger: Nasopraemaxilla
Lat:
Rus:
Sp: naso-premaxila

Name given by Mayhew (1924) to the premaxilla of *Lepisosteus* and *Amia*, because it is formed by the fusion of a dentigerous bone of dermal origin, the *premaxilla,* with one or two pairs of bony laminar wings, the *nasals.*

231. Neural arch

Fr: arc neural
Ger: Neuralbogen
Lat: arcus neuralis (pl: arcus neurales); arcus superior (pl: arcus superiores)
Rus: невральная дуга (pl: н-ые дуги)
Sp: arco neural

Arch formed in the embryo by the basidorsals and, more rarely, by the interdorsal elements of the vertebrae. In adults, the branches forming the neural arch are called *neurapophyses.* In chondrosteans, the neural arch is, in general, a continuous structure. In teleosts, the arches are separate, thus giving more flexibility to the vertebral column. The neural arches with their spines are directed backwards. In sturgeon (*Acipenser*) the neural arches form two canals, one for the spinal cord and another above for the *longitudinal ligament.*

See   Vertebrae
Fig: 35 C

232. Neural canal

Fr: canal neural
Ger: Neuralkanal
Lat: canalis neuralis (pl: canales neurales)
Rus: невральный канал (pl: н-ые каналы)
Sp: canal neural

Space formed by the neural arch, through which the spinal cord runs.

Fig: 35 C

233. Neural spine

Fr: épine neurale
Ger: Neuralfortsatz
Lat: processus spinosus dorsalis (pl: processus spinosi dorsales)
Rus: невральная игла (pl: н-ые иглы)
Sp: espina neural

A thin, long bone formed above the neural arch to which it later fuses, resulting in the characteristic spines of the teleost backbone. Diodontidae have bifid neural spines in the posterior section of the vertebral column.

Syn: neuracanth; neurapophysis (pl: neurapophyses)
Figs: 13 and 35 C

234. Neurocranium

Fr: neurocrâne
Ger: Hirnschädel; Neurocranium

Lat: neurocranium (pl: neurocrania)
Rus: неврокра́ниум; мозговóй чéреп (pl: мозговы́е чéрепы)
Sp: neurocráneo

The neurocranium or braincase is comprised of the cartilages and bones that provide direct support and protection to the brain and to the three most important sense organs, smell, sight, and hearing. The neurocranium proper was the earliest component of the skull.

In modern forms, its embryonic development follows a common pattern, applicable to all the remaining vertebrates, including man. Two pairs of cartilaginous bars appear at its base, one anterior (*prechordal* or *trabecular bars*) and one posterior (*parachordal bars*). These are later joined by three transverse plates (*commissures*). This assemblage grows on the sides and back forming a box, the *neurocranium* proper, with numerous openings of variable size to allow the passage of the notochord, nerves and blood vessels.

To these first units of the braincase are soon added three pairs of capsules : the *olfactory capsules* in front, followed by the *optic, orbital* or *sphenotic capsules,* and posteriorly, the *otic* or *auditory capsules*.

All the bones formed by ossification of this cartilaginous template are called chondral or endochondral bones. Their number varies from group to group, although there is a recognizable common pattern. This variety of ossifications, followed by fusion and splitting of previous bones, make the comparison among the fish species very difficult, even when following their embryonic development. As a consequence, the terminology of fish bones is subject to continuous revisions and debates.

The neurocranium bones leave spaces in the areas of contact thereby allowing skull growth. These cartilages disappear in old individuals of the most evolved bony fishes.

The bones derived from the neurocranium are of chondral origin and can be grouped into the following five regions, listed below with their most common bones:

Region	Bones
1. Ethmoidal	1.1 Ethmoid
	1.2 Lateral ethmoid
2. Orbital	2.1 Orbitosphenoid
	2.2 Pterosphenoid
	2.3 Basisphenoid
3. Otic	3.1 Prootic
	3.2 Opisthotic
	3.3 Autopterotic
4. Occipital	4.1 Epioccipital
	4.2 Supraoccipital
	4.3 Exoccipital
	4.4 Basioccipital

To all these bones of the *neurocranium* proper, which are kept solidly together by complex sutures, some bones of the dermocranium (prevomer, frontals, parietals, parasphenoid, and intercalar) have been intimately added, in such a way that they form a unit in advanced actinopterygian fishes.

In the preparation of skulls and sometimes in the digs, this complex is found in one piece. This unit is also referred to in some biological and archaeological works as *neurocranium*. Only in a *sensu lato*, however, can this second interpretation be accepted.

Fig: 2 and 14 A
Bibl: Beer (1937)

235. Notochord

Fr: notochorde
Ger: Chorda
Lat: chorda dorsalis (pl: chordæ dorsales)
Rus: нотохо̄рда; хо̄рда; спиннāя струнā
Sp: notocordio

Although the notochord will not likely be found in archaeological sites, its importance in the evolution of fishes and tetrapods is so great that it deserves to be studied here.

The term *notochord* is applied to the following three biological entities:

a)   one of the five original embryonic layers (the others being the ectoderm, endoderm, mesoderm and neurectoderm). From these five embryonic tissues all the organs of chordate embryos develop;

b)   a special skeletal tissue, typical of chordate embryos and lower fishes, characterized by the presence of large cylindrical cells arranged in columns, like a pile of coins, with their nuclei off-center. As the cells mature, their vacuoles enlarge and fill with fluid, contributing to its typical turgescence; and

c)   a permanent or transitory organ found in the chordates, that developing at the back of the embryo extends from the tail to the level of the hypophysis in the brain. It lies immediately below the spinal cord. Its posterior end is pointed while the anterior end is round. The notochord, as an organ, has a triple function. It supports the delicate spinal cord; it serves as an attachment for the body musculature, and it prevents the animal from telescoping during swimming movements and contractions of the body.

Evolutionary significance of the notochord

The notochord represents the oldest skeletal organ of chordates. Its presence in embryos and adults is so characteristic of this taxonomic group, which includes a variety of forms, from Hemichordates to man, that it gave the group its name.

The notochord acts as the foundation of the vertebral column by either substitution or transformation of the notochordal tissue and its sheaths into bony vertebrae. This change was carried out very slowly through an evolutionary process beginning with the fossil agnathans and placoderms, all of which possessed this axial skeletal unit.

Cyclostomes are the first fishes to have the rudiments of a vertebral column, with the formation of the neural and hemal arches, although the notochord remains unchanged for life.

In elasmobranchs, the skeletogenous tissue is found outside of the notochordal sheaths, but it migrates into the center of the notochord reducing its thickness and segmenting it into cylindrical units. This reduction is due to the invasion of calcium salts that penetrate the sheaths and sometimes the inner cylinder. The calcification occurs in the form of concentric or radial bands, producing the different types of elasmobranch vertebrae, known as cyclospondylous, asterospondylous and tectospondylous. In holocephalans, the notochord is cylindrical and is covered with a cartilaginous sheath, without vertebral centra.

In <u>actinopterygians</u>, the skeletogenous tissue that forms the neural and hemal arches remains outside of the notochordal sheath. In <u>chondrosteans</u> (paddlefishes and sturgeons), the notochord remains as a cylinder without constrictions and can attain a length of three and a half meters with a diameter of a little more than one centimeter. In these fishes the neural and hemal arches are well developed but remain cartilaginous. In most <u>holosteans</u>, the notochord persists in the adult, while in some, the vertebral column is completely developed (*Amia* and *Lepisosteus*).

In <u>teleosts</u>, the notochord reduces its size until finally it remains as a soft tissue filling the anterior and posterior cavities of the vertebrae. The remnant of the notochord forms a continuum, since the masses that fill the cavities join by means of a thread-like tissue that runs along a narrow canal in the vertebrae. This arrangement gives the teleostean notochord a *moniliform* appearance, e.g., as a string of pearls. The reduction in volume of the notochord is accompanied by a reduction in its mass, owing to a degeneration and vacuolization of the notochord.

236. Occipital crest

Fr: crête occipitale
Ger: Occipitalleiste (= Okzipitalleiste); Hinterhauptskamm
Lat: crista occipitalis (pl: cristæ occipitales)
Rus: затылочный гребешóк (pl: затылочные гребешóк)
Sp: cresta occipital

The occipital bone of teleosts carries a middle vertical lamina formed by the ossification of the connective septum that separates the right and left bundles of myomeres in the occipital region. The occipital crest becomes especially large in those fishes whose body is deeper than the head.

Fig: 14 B

237. Occipital region

Fr: région occipitale
Ger: Occipitalregion; Hinterhaupstregion
Lat: regio occipitalis
Rus: затылочная óбласть
Sp: región occipital

The occipital region forms the most posterior part of the neurocranium. This region connects with the vertebral column by means of three facets, two lateral, each one corresponding to an exoccipital, and one median, corresponding to the basioccipital.

The occipital region fuses during embryogeny with the otic region, and together they form a unit that encloses the membraneous labyrinth and the largest part of the brain. The bones of the occipital region are either of a chondral origin or are formed from a combination of both types, chondral and dermal. According to this criterion the bones can be grouped into two categories, noting nevertheless that not all bones listed are present in all fish groups.

Endochondral	*Mixed*
exoccipital (2)	dermosupraoccipital (1)
basioccipital (1)	supraoccipital (1)
occipital (1)	

Bibl: Shute (1972)

238. Opercle   *

    Fr: opercule; operculaire
    Ger: Operculum
    Lat: os operculum (pl: ossa opercula); os operculare (pl: ossa opercularia)
    Rus: оперкулярная кость   (pl: оперкулярные кости)
    Sp: opérculo; opercular

Paired bone of dermal origin that reinforces the dorsal part of the opercular membrane. It is usually the largest bone of the opercular series. It is acknowledged that it derives from one of the branchiostegal rays of primitive bony fishes.

The opercle very often displays growth lines, a feature that has been used to estimate the age of the fish from some families (Percidae).

    Syn: Opercular; operculum
    Figs: 1 C, 3, 6, 13, 14 B, and 23
    Bibl: Rojo (1986); Fagade (1974)

239. Opercular membrane

    Fr: membrane operculaire
    Ger: Opercularmembran; Hautlappen
    Lat: membrana branchialis (pl: membranæ branchiales)
    Rus: жаберная покрышка (=мембрана); крышечка
           (pl: жаберные покрышки (=мембраны); (pl: крышечки)
    Sp: membrana opercular

The opercular membrane, is an integumental flap, typical of bony fishes, that covers the gill clefts and it is reinforced by four dermal bones: opercle, preopercle, subopercle and interopercle. The assemblage of the membrane and the bones receives the name of *operculum*. By consensus, it is admitted that the fish head extends back to the border of this membrane.

This membrane is fixed to the cheek on its anterior border, with the preopercular bone acting as a hinge. The posterior border is free allowing the water entering the mouth and bathing the gill membranes to escape. Sometimes there is a membranous cover overhanging the bony border of the operculum. The opercular cleft is wide in most cases, but it can be reduced to a small circular opening, as in *Anguilla*. When both opercular membranes converge on the isthmus, the left side overlaps the right operculum in the majority of the Teleosts.

    Syn: operculum
    Fig: 6
    Bibl: Hubbs (1919); Nelson (1949)

240. Opisthocelous vertebra   *

    Fr: vertèbre opisthocœlique
    Ger: opisthocoeler Wirbel
    Lat: vertebra opisthocœlica (pl: vertebræ opisthocœlicæ)
    Rus: опистоцельный позвонок (pl: опистоцельные позвонки)
    Sp: vértebra opistocélica

As its name implies, the posterior face of the centrum of the opisthocelous vertebra is concave. Among all fishes, the opisthocelous vertebra is only found in the gars (*Lepisosteus*). This feature makes this type of vertebra easily recognizable.

241. Opisthotic

Fr: opisthotique
Ger: Opisthoticum
Lat: os opisthoticum (pl: ossa opisthotica)
Rus: заднеу́шая кость (pl: заднеу́шие ко́сти)
Sp: opistótico

Name proposed by Huxley (1858) for the posterior ossification of the otic capsule of actinopterygians. The opisthotic covers the ampulla of the posterior semicircular canal.

Vrolik (1873) named this bone *intercalar* when he demonstrated that it is not related to the labyrinth and it is not an endochondral bone. The true endochondral opisthotic is, apparently, present in fossil crossopterygians and fossil actinopterygians.

Since the true opisthotic is lost in *Amia, Lepisosteus,* and modern teleosteans, more and more biologists use the name *intercalar* to replace the older term *opisthotic,* which still is used extensively.

See Intercalar

242. Orbital region

Fr: région orbitale
Ger: Orbitalregion
Lat: regio orbitalis
Rus: глазна́я о́бласть
Sp: región orbitaria

The region of the fish head around the eye, covered by bones of dermal origin, which may or may not be related to the sensory canals. It is a region of great plasticity, the number of bones increasing as the rostrum or the cheek areas lengthen. The circumorbital bones can be divided into two categories, as indicated below.

Canal bones	*Bones not related to sensory canals*
antorbital	suborbitals (several pairs)
infraorbitals (1-6 pairs)	supraorbitals (SPO1 and SPO2)
[lacrymal (IO$_1$)	
jugal (IO$_2$)	
postorbitals (IO$_3$, IO$_4$ and IO$_5$)	
dermosphenotic (IO$_6$)]	

243. Orbitosphenoid *

Fr: orbitosphénoïde
Ger: Orbitosphenoid
Lat: os orbitosphenoideum (pl: ossa orbitosphenoidea)
Rus: глазни́чная клиновидная кость; орбитосфенбид

Sp: orbitoesfenoides

Median or paired bone of endochondral origin formed in the septum that separates both orbits. It forms part of the floor and walls of the cranial cavity. It articulates with the lateral ethmoid in front, and posteriorly with the pterosphenoid. It is missing in Salmonidae, and Gadidae, among other fish families.

Fig: 14 A

244. Ostariophysans

Fr: Ostariophysaires; Ostariophyses
Ger: Ostariophysier
Lat: Ostariophysi
Rus: костнопузырные рыбы
Sp: Ostariofisos

Taxonomic group established by Sagemehl (1885) to include those bony fishes provided with Weberian apparatus. The ostariophysans make up a very natural and uniform group that takes its name from the presence of a chain of four ossicles: *claustrum, scaphium, intercalarium* and *tripus,* which connect the otic capsule to the gas bladder. After the studies of Bridge and Haddon (1889), it has been proven that these bones, although related to the ear capsule, are not homologous with the mammalian ear bones.

Following Berg (1940) the ostariophysans have been renamed Cypriniformes, which make up the most numerous order among the freshwater fishes, having some 5,000 species (Darlington, 1957).

Table   2

245. Osteichthyans

Fr: Ostéichthyens
Ger: Knochenfische
Lat: Osteichthyes
Rus: кóстные рыбы
Sp: Osteictios

The osteichthyans or bony fishes, so named because of the ossification of their skeleton to a greater or lesser degree, comprise the third class of FISHES. They are the most numerous in species (90 % of extant fishes), they have adapted to more varied conditions in the aquatic environment and they have become the most specialized in their anatomy and physiology.

The most important osteological features of this group are the presence of
a)   mandibles with teeth,
b)   paired fins with their corresponding girdles,
c)   an ossified skeleton,
d)   complete vertebrae in most species,
e)   branchial arches with an internal bony support,
f)   branchial arches not connected with the skull, and
g)   overlapping scales that cover most of the body in the majority of the species.

The osteichthyans form a natural group, recognized for the first time by Willughby (1686) in his *Historia piscium*. On the basis of a combination of different criteria, the group is divided into three subclasses: actinopterygians, crossopterygians and dipnoans. Only the actinopterygians are represented in North American fresh- and salt waters.

Syn: Bony fishes
Tables 2 and 3

246. Osteocranium

Fr: osteocrâne
Ger: Knochenschädel; Osteocranium
Lat: osteocranium (pl: osteocrania)
Rus: кóстный чéреп
Sp: osteocráneo

The assemblage of bones that, at any given stage of fish development, are present in the skull of an osteichthyan fish. This term is the counterpart of chondrocranium (see this term). As the chondrocranium (which appears first in the fish embryo) decreases in extension during the embryonic development of the fish, the osteocranium increases in size and in the number of its elements.

247. Otic region

Fr: région otique
Ger: Otikalregion; Ohrenregion
Lat: regio otica
Rus: ушнáя óбласть
Sp: región ótica

That region of the skull containing the hearing and equilibrium organs and the bones associated with them. This region is located between the orbitosphenoid and the occipital, to which it is solidly fused. The otic region comprises bones of endochondral, dermal and mixed origin, as shown below:

Endochondral		_Dermal_	_Mixed_
autosphenotic (2)		intercalary (2)	sphenotic (2)
autopterotic (2)	—	pterotic (2)	prootic (2)
opisthotic (2)			
epiotic (2) *			

*According to Patterson (1975), the epiotic belongs phylogenetically to the occipital region.

See   Epiotic.

248. Otoliths   *

Fr: otolithes
Ger: Otolithen (sing: Otolith); Gehörsteine
Lat: otolithi (sing: otolithus)

Rus: ОТОЛЙТЫ   (sing: ОТОЛЙТ)
Sp: otolitos

The *otoliths, statoliths* or *"ear-stones"* are hard concretions of varying size and chemical composition, found in the otic capsule of ray-finned fishes. They are enclosed in the chambers of the inner ear: the *lapillus* is lodged in the utricle, the *sagitta*, in the saccule, and the *asteriscus,* in the lagena. Most of them are made of calcium carbonate, crystallized in the form of aragonite. The otoliths are attached to the chamber walls by ligamental formations, called *marginaria* and play an important role in maintaining equilibrium, and possibly in hearing.

Although the otoliths of fishes are not bones, they are usually included in the study of the skeleton, because of their mineralized nature and their resistance to environmental destructive factors, after the death of the fish.

Cartilaginous fishes (sharks and rays) do not have otoliths; instead they have calcareous granules, *otoconia* (sing: *otoconium*), mixed with mineral particles, *otarenae* (sing: *otarena*), that have entered through the endolymphatic pore. They are too small to be detected in normal archaeological work.

Chondrosteans have otoliths made of *vaterite*, a very unstable form of calcium carbonate. Sturgeons, which belong to this group, have both otoliths and otoconia made of crystals of aragonite. The holosteans have otoliths composed of a mixture of vaterite and aragonite, while teleosts have otoliths made of aragonite.

Otoliths grow by the superposition of calcium carbonate layers interspersed with a protein, *conchioline*, which is related to the organic material of the shells of certain molluscs. When cut, otoliths display an alternation of bands or zones. When viewed by transmitted light, they appear as alternating hyaline and opaque bands, but if viewed by direct light over a black background, these bands show up as dark and clear, respectively. A hyaline and opaque band together form an annulus or ring, which represents the growth of an entire year. The term *annulus*, is also applied in a stricter sense to the *winter zone*. These bands are also called *summer* and *winter* zones, being very wide in the first case and narrow and compact in the second. The thickness of each band results from the growth rate of the otolith in each period, reflecting the abundance of food in summer and its scarcity in winter. Here, "summer" refers to the whole growth season, which in some cases can extend from spring to fall, and "winter", to the period of no growth, restricted to the climatic winter and the beginning of spring.

This regular arrangement of the bands has been used to estimate the age of many species of fishes. The reading of the annuli is made using a stereoscopic microscope. Thin otoliths can be viewed directly, while thick ones can be prepared by burning, staining or cutting. A word of caution should be added here, since there are false rings, called *checks* or *splits* , resulting from several factors (spawning, diseases, migrations, lack of food, changes in temperature, etc.) and which should not be confused with yearly rings. Recent studies (Panella, 1971) have shown that otoliths are built up of daily deposition of layers of inorganic and organic matter.

The morphology of the otoliths is different in each species, but similar within a given family. At the same time, their mineral composition favours the production of remarkable fossils that resist environmental deterioration for a considerable time. As a result of this set of factors, they are easily recognized and are used in systematics to establish fish relationships; in paleontology, to recognize the relationship of fossil forms with extant fishes; in paleogeography, to trace the boundaries of ancient seas; in archaeology, to identify fish utilized in prehistoric human settlements; and in biology, for age determination and to study the fish diet of marine animals.

Within the same species, there is a direct relationship between otolith size (length and weight) and fish size, which has been used successfully as a discriminatory factor for the recognition of codfish populations (Rojo, 1977). The size of the otolith is not related to the size of the fish species in a general sense, since there are species of large-size fishes (*Salmon, Anarhichas*) which have minute otoliths, and species of small-size fishes (*Aplodinotus*), with very large ones. But within a single species, the size of the otolith is directly proportional to the size of the fish. In most species, the largest otolith is the sagitta, but in Cypriniformes, the largest otolith is either the asterisk or the lapillus, never the sagitta.

Uses of the otoliths

Otoliths are very valuable in biological and archaeological studies because they can provide, directly or indirectly, a large amount of biological information. Their morphological features make possible the identification of the fish family, genus and species. The characteristics of the otolith that can have an identification value are: thickness; smoothness and rugosity of its faces; the outline of its edges, which can be smooth, scalloped, serrate, or notched; the shape and length of the acoustic groove; and the features of the ostium and cauda.

Once the species has been identified, the physical characteristics of the environment can be deduced (water depth and temperature, climatic conditions). Their dimensions (maximum length, width and weight) are used to estimate the live size of the fish, and if a large collection of otoliths is available, the growth rate and size range of the fish can also be estimated. Numerous biological papers offer regression coefficients between otolith length or weight and fish live size. The closer in space and time the biological sample is to the archaeological one, the more valuable these coefficients are. If no regression equations are available for a given fish in a given place, it is easy to calculate one, using a sample taken from present fish from the closest area.

One of the most important applications of otoliths is the estimation of the fish age and the time of its capture. As a consequence, the otolith provides an indirect clue to the seasonality at the site. In fact, otoliths have proved to be more reliable than scales for age estimation. The otoliths used in these cases are the sagitta for most fishes or the asterisk and the lapillus (whichever is the largest) for Cypriniformes. Thin otoliths (Pleuronectiformes) are ground down to reduce their thickness before being studied under the microscope. A magnification of 20 to 30 diameters is sufficient in most cases. Thick otoliths (Gadiformes) are cut across at the level of the collum, which corresponds to the core of the otolith. In this way it is possible to see all growth layers. In both cases, the growth annuli are best seen by placing a drop of glycerine on the cut surface and by directing a screened beam of light on the otolith. This is done by placing a file card vertically at the level of the cut surface, between the otolith and the source of light.

The most important families from which otoliths have been used to estimate the age of the fish are: Anguillidae, Clupeidae, Engraulidae, Batrachoididae, Gadidae, Atherinidae, Malacanthidae, Haemulidae, Scombridae, Bothidae, Pleuronectidae, and Carangidae.

As explained earlier, the cyclic growth of the otolith allows the evaluation of fish growth in the year in which it was captured. The width of the outermost layer in the otolith, compared to the previous layer, indicates the season of capture. When both widths are very similar, the fish, most likely, was caught at the end of the growing season or during the winter, when growth stops. If the outer layer is much narrower than the preceding one, it corresponds to a spring, summer or fall capture. The thickness of this outer layer is proportional to the part of the year that has elapsed since the beginning of the growing season. In this case, the fish has not had enough time to complete the annual growth of that year.

Since the lapillus and the asterisk are very small in most cases, it is best to utilize a wet-sieving method (10-30 mm mesh) as recommended by some archaeologists to recover them from the dig.

Otoliths or "earstones" have been used by some aboriginal peoples as ornaments, due to their large size or appealing shape (Niehoff, 1952).

Figs: 42, 43, 44, and 45

Bibl: Adams (1940); Casteel (1974 b; 1976); Degens, Deuser, and Haedrich (1969); Fitch (1958); Gulland (1958); Niehoff (1952); Panella (1971); Pradhan and Kapadia (1953); Priegel (1963); Rojo (1977); Scott (1906); Tatara, Yamaguchi and Hayashi (1962); Templeman and Squires (1956); Witt (1960).

249. Paired fins

Fr: nageoires paires
Ger: paarige Flossen
Lat: pinnæ pariles or pares; artiopterygium (pl: artiopterygia)
Rus: па́рные плавники́
Sp: aletas pares

The paired fins are located in a symmetrical position on the fish body, the pectoral fins in an anterior position and the pelvic fins usually placed behind them. Although the pectoral and pelvic fins are sometimes different in structure, they are considered homologous. The number of paired fins is never more than four, but often this number is reduced to two, due to the absence of the pelvic fins, as in apodes fishes.

Syn: artiopterygia
Fig: 6
Bibl: Eaton (1945); Goodrich (1906)

250. Palatine   *

Fr: palatin
Ger: Gaumenbein; Palatinum
Lat: os palatinum (pl: ossa palatina)
Rus: нёбная (=палати́нная) кость [pl: н-ные (=п-ные) ко́сти)]
Sp: palatino

In modern teleosts, the palatine is a bone of mixed origin resulting from two ossifications: one endochondral, on the anterior part of the palatoquadrate, the *autopalatine*, and the other dermal, the *dermopalatine*.
The autopalatine is present in primitive actinopterygians with the exception of *Lepisosteus*, in which it remains as cartilage. The dermopalatine remains independent, as in *Amia*, or fuses to the autopalatine, as in most teleosts. In some cases teeth are later added to the dermopalatine. The presence of teeth in this bone is an apomorphic character.
    —   In spite of the double nature of this bone, it is customary to call it *palatine*.

See   Autopalatine and Dermopalatine
Figs: 1 D, 13, 25 A, and 27

251. Palatoquadrate bar

Fr: palatocarré
Ger: Palatoquadratum
Lat:
Rus: нёбноквадра́тный хрящ
Sp: palatocuadrado

Cartilaginous bar forming the upper mandible of the embryos of gnathostome fishes. It remains cartilaginous during the whole life of chondrichthyan fishes.

Both palatoquadrate bars meet in the middle line, forming the mandibular symphysis, which allows them to move slightly. The palatoquadrate bars and Meckel's cartilages together form the mandibular arch, which according to modern interpretation, has evolved from the first (or one of the first) branchial arches of early fishes.

In adult actinopterygians, dipnoans and crossopterygians, the palatoquadrate cartilage presents three ossifications, producing the following endochondral bones: the autopalatine, the metapterygoid, and the quadrate.

Since cyclostome fishes lack mandibles, they obviously do not possess these bars.

Syn: palatoquadrate cartilage; pterygoquadrate bar; epimandibular cartilage; maxillar cartilage.

252. Parachondral bones

Fr: os parachondrals
Ger: parachondrale Knochen
Lat:
Rus: парахондра́льные ко́сти (sing: парахондра́льная кость)
Sp: huesos paracondrales

Bones derived from the ossification of the connective tissue that surrounds a cartilage. The ossification process extends later to the subjacent cartilage, producing a bone of mixed origin, but predominantly of chondral origin. This type of ossification is common in bones formed in the sensory capsules. Many authors call them chondral or endochondral bones.

253. Parallelism

Fr:
Ger: Parallelismus
Lat:
Rus: параллели́зм
Sp: paralelismo

The similarity of shape and structure involving organisms which have a common ancestry, when the similarity is the result of environmental adaptations.

254. Parapophysis

Fr: parapophyse
Ger: Querfortsatz, Parapophyse
Lat: parapophysis (pl: parapophyses)
Rus: парапо́физ (pl:парапо́физы); боково́й отро́сток
Sp: parapófisis

Strictly speaking, the parapophysis is the vertebral process that connects the body of a vertebra to the *capitulum* of a rib in mammals. In fish anatomy, this term is applied to the long, lateral processes that protrude from the vertebral centra of the abdominal region of the vertebral column. The parapophyses support the epipleural bones (dorsal ribs) when present, and in other cases the gas bladder, as in Gadidae. In Clupeiformes, the parapophyses are not fused to the vertebrae.

Syn: transverse processes; basapophysis
Fig: 35 C

255. Parasphenoid   *

Fr: parasphénoïde
Ger: Parasphenoid
Lat: os parasphenoideum (pl: ossa parasphenoidea)
Rus: парасфеноидная кость; парасфеноид
Sp: paraesfenoides

Median bone of dermal origin that forms the base of the skull of bony fishes, although it tends to disappear in modern teleosts. Two types of parasphenoids are known: one short, and the other extending from the vomer in the ethmoidal region to the basioccipital in the occipital region. In elopomorph, clupeomorph, and osteoglossomorph fishes, the parasphenoid has sharp little teeth (*Amia*) or molariform teeth (Albulidae). The parasphenoid presents two ascending processes in most teleosts. According to Gregory (1933), it is homologous with the mammalian vomer. Authors who follow this interpretation call the vomer *prevomer*. With this in mind, the terms *vomer* and *prevomer* in osteological works have to be considered synonymous.

Figs: 13, 14 B, and 20

256. Parhypural

Fr: parhypural
Ger: Parhypural
Lat:
Rus: паргипуралия
Sp: parahipural

The modified hemal spine of the first preural centrum. It is the last hemal arch crossed by the dorsal aorta. In old works, it was considered erroneously as a hypural, error still maintained in some modern papers.

Fig: 37

257. Parietal   *

Fr: pariétal (pl: pariétaux)
Ger: Parietale
Lat: os parietale (pl: ossa parietalia)
Rus: теменная (= париетальная) кость
Sp: parietal

Laminar dermal paired bone that covers the otic region in bony fishes. When both parietals meet in the middle line of the skull, the frontal and the supraoccipital are kept apart. This type of skull is called *medioparietal* xand is found in *Amia,* and in the family Elopidae among other fishes. In the skull of the type *lateroparietal*, the frontals and the supraoccipital meet in the center, as in Gadidae, leaving one parietal on each side of the skull. The parietal is absent in Syngnathiformes, Siluridae, etc., hence this type of skull is called *aparietal*. In the latter case, it is accepted, that the parietals have fused with the supraoccipital, giving origin to the bone called *parietooccipital*.

Figs: 2, 13, and 14 B

258. Parietooccipital

Fr: pariéto-occipital
Ger: Parietooccipitale
Lat: os parietooccipitale (pl: ossa parietooccipitalia)
Rus: затылочно-теменная кость
Sp: parieto-occipital

Median dermal bone located on the dorsal side of the skull formed by the fusion of the parietals with the supraoccipital, as in Siluridae.

See   Parietal

259. Pavement teeth

Fr: dents en pavé
Ger: Pflasterzähne
Lat:
Rus:
Sp: dientes en pavimento

Some fishes feeding on molluscs have their teeth packed together in large plates, formed by rows of teeth of identical or different size and shape, with which they can easily crush the shells of mollusks and crustacea. This type of arrangement is known as pavement or mosaic teeth. Among sharks, only the genera *Mustelus, Hexanchus,* and *Heterodontus* have pavement teeth.

260. Pectoral fins

Fr: nageoires pectorales
Ger: Brustflossen
Lat: pinnæ pectorales (sing: pinna pectoralis); omopterygium (pl: omopterygia)
Rus: грудные плавники (sing: грудный плавник)
Sp: aletas pectorales

Paired fins located in the thoracic region (*omopterygia* [sing: *homopterygium*]) homologous to the anterior limbs of the tetrapods. These fins are generally larger and have more rays than their corresponding pelvic fins. They are also more stable in their structure and position than pelvic fins, which have evolved more rapidly.

In more primitive bony fishes, the pectoral fins are placed immediately behind the head and set very low on the fish flank. In more advanced bony fishes, they are set high on the side, with the pelvic fins (when present) immediately below or in front of them.

See   Paired fins
Figs: 6 and 13

261. Pectoral girdle

Fr: ceinture scapulaire
Ger: Schultergürtel
Lat: cingulum pectorale (pl: cingula pectoralia)
Rus: плечевой пояс
Sp: cintura pectoral; cintura escapular

The pectoral fins of both cartilaginous and bony fishes are attached to the skeleton by an ensemble of cartilages and bones, respectively, known as the pectoral girdle.

In bony fishes, the bones forming the pectoral girdle have a double origin (Sewertzoff, 1926). The coracoid and the scapula, of endochondral origin, form the *primary pectoral girdle*. In some cases, there is a *mesocoracoid* between these bones. A second group of bones (posttemporal, supracleithrum, cleithrum and from one to three postcleithra) derived from the dermoskeleton of the cephalothorax, forms the *secondary pectoral girdle*. In some cases, a mesocleithrum and a hypocleithrum are also present.

The terms primary and secondary refer to the relationship between the girdle and the pectoral fins. The primary girdle directly supports the radials and fins, while the secondary is only indirectly related to the fins. Although the connexion of the fins to the skull by the posttemporal was attained later in evolution, the bones of the secondary girdle are phylogenetically older than those of the primary girdle.

> Syn: scapular girdle
> Figs: 1 E, 13, 25 B, and 38
> Table   4

262. Pelvic fins

> Fr: nageoires pelviennes
> Ger: Bauchflossen
> Lat: pinnæ ventrales or abdominales (sing: pinna ventralis or abdominalis); ischiopterygium (pl: ischiopterygia)
> Rus: брюшные плавники (sing: брюшный плавник)
> Sp: aletas pelvianas; aletas pélvicas

The paired fins which are located originally in an abdominal position (*ischiopterygia* [sing: *ischiopterygium*]) and are homologous with the posterior appendages of the tetrapods. The number of rays is reduced in comparison with the number of rays found in the pectoral fins.

The pelvic fins have been exposed to more structural and functional changes in the evolution of fishes, than their corresponding pectoral fins. In bony fishes the pelvic fins are located in the lower part of the fish flank, but at a different relative distance from the head depending on the group phylogeny. This character was accepted by Linnaeus to classify bony fishes into four groups: *abdominal fishes*, with the fins far back in the abdomen (*Salmo*); *thoracic fishes,* if the fins have moved under the chest region; *jugular fishes,* if found under the throat or the chin, and, lastly, *apode fishes* , that lack pelvic fins altogether. This classification, well accepted during the whole of the last century, has been replaced by modern classifications based on several anatomical characters taken together.

These fins are sometimes modified to perform special functions related to the reproductory activity. In other cases the pelvic fin bear a stout spine that sticks outward making it difficult for a predator to swallow the fish, as it happens in the sticklebacks.

> See   Myxopterygium; Ventral fins
> Figs: 6 and 13

263. Pelvic girdle

> Fr: ceinture pelvienne
> Ger: Beckengürtel

Lat: basipterygium (pl: basipterygia)
Rus: брюшнóй (= тáзовый) пóяс
Sp: cintura pelviana (= pélvica)

The pelvic fins of Chondrichthyes and Osteichthyes are supported respectively by cartilages and bones known collectively as the *pelvic girdle*. The pelvic girdle is always simpler in structure than its corresponding pectoral girdle and is never connected to the vertebral column.

In bony fishes, the pelvic girdle is exposed to greater structural and positional modifications than those affecting the pectoral girdle. In holosteans (*Acipenser*), the pelvic girdle is formed by the fusion of elements of endochondral origin. In holosteans (*Amia*) and teleosts, it is composed of two bones, usually triangular in shape, known as the pelvic bones (*basipterygia*).

In thoracic and jugular fishes, the pelvic girdle and its fins migrate forward to articulate with the pectoral girdle. Apode fishes have lost both the pelvic girdle and pelvic fin.

See Basipterygium
Figs: 1 E, 13, and 41

264. Perichondral bones

Fr: os périchondrals
Ger: perichondrale Knochen
Lat:
Rus: перихондрáльные кóсти (sing: перихондрáльная кость)
Sp: huesos pericondrales

The ossification that produces a perichondral bone starts at the perichondrium of a cartilage and extends rapidly to the inner cartilaginous mass. Perichondral bones are sometimes also simply called chondral or endochondral bones.

Table 1

265. Perichondrium

Fr: périchondre
Ger: Knorpelhaut; Knorpelhülle; Perichondrium
Lat: perichondrium (pl: perichondria)
Rus: надхрящница; перихóндрий
Sp: pericondrio

A sheath of connective tissue that covers a cartilage.

266. Periosteum

Fr: périoste
Ger: Knochenhaut; Periost
Lat: periosteum
Rus: периóст; надкóстница
Sp: periostio

A sheath of connective tissue that covers a bone.

267. Pharyngeal process

Fr: processus pharyngien
Ger: Schlundfortsatz; Pharyngelfortsatz
Lat:
Rus: глóточный отрóсток (pl: глóточные отрóстки)
Sp: proceso faríngeo

Large posterior expansion of the basioccipital of cyprinoid fishes. It is located above the pharyngeal toothplates of the fifth ceratobranchials. Its specific shape is a good identification feature.

Figs: 36 A

268. Pharyngobranchials   *

Fr: pharyngobranchiaux (sing: pharyngobranchial)
Ger: Pharyngobranchialia
Lat: 1. cartilagines pharyngobranchiales (sing: cartilago pharyngobranchialis)
        2. ossa pharyngobranchialia (sing: os pharyngobranchiale)
Rus: 1. фарингобранхиáльные хрящú (sing: ф-ный хрящ)
        2. фарингобранхиáльние кóсти (sing: ф-ная кость)
Sp: faringobranchiales

Endochondral bones that make up the uppermost segment of the branchial arch skeleton. In some primitive actinopterygians and crossopterygians there were two pharyngobranchials, for which Van Wijhe (1882) proposed the terms, *suprapharyngobranchials* , never associated with teeth and the *infrapharyngobranchials*, sometimes associated with toothed dermal plates. The term infrapharyngobranchials refers to their position below the suprapharyngobranchials, but both are located in the dorsal area of the pharyngeal chamber. One is not likely to find suprapharyngobranchials in archaeological sites, since these bones are only present in very primitive fishes. Since the term infrapharyngobranchial, though correct, is cumbersome, most authors name it simply *pharyngobranchial,* be it associated with a tooth plate or not.

Figs: 4 and 25 A

269. Placoderms

Fr: Placodermes
Ger: Panzerfische; Placodermen
Lat: Placodermi
Rus: Плакодéрмы
Sp: Placodermos

Class of fishes represented only by fossil forms, which appeared in the upper Silurian and lower Devonian. Their dominance was of limited duration, since they disappeared almost completely by the end of the Devonian period. The name refers to the dermal bones, which in the form of plates covered the head and thorax. In certain cases the rest of the body was also covered by smaller plates.
Placoderms are the first fishes having mandibles and a bony skeleton, and are more closely related to modern chondrichthyans than to any other living group.

Table   4

270. Placoid fishes

Fr: poissons placoïdes
Ger: Placoidfische (= Plakoidfische)
Lat: pisces placoidei
Rus: плакоидные рыбы (sing: плакоидная рыба)
Sp: peces placoideos

Agassiz (1833) classified fishes into four groups according to the structure of their scales. The first group, placoid fishes, included those fishes covered with placoid scales or denticles, such as sharks, rays, skates, and chimeroids.

Table   2 and 4

271. Placoid scales   *

Fr: écailles placoïdes
Ger: Placoidschuppen (= Plakoidschuppen)
Lat: pisces placoidei
Rus: плакоидные чешуи (sing: плакоидная чешуя)
Sp: peces placoideos

Modern placoid scales evolved from the thick cosmoid scales of placoderms, through the loss of the deep layers of bony and spongy tissue, and the retention of the outer layer of dentine only. They are also called *denticles,* because of their shape and structure, and *dermal denticles,* because they derive from the dermis.

Placoid scales consist of a basal plate of quadrangular or stellate shape embedded in the dermis, from which arise a curved cone or denticle directed backwards. The hollow center of the cone, is called the *pulp cavity,* because it is filled with soft material: blood vessels, nerves and specialized dermic cells, the *odontoblasts,* that produce dentine or a similar material that makes up the body of the scale. The outer layer of the denticle consists of a special type of enamel, *vitrodentine,* secreted by the epidermal cells, the *ameloblasts.*

Placoid scales can resist well erosion and decomposition. Denticles of sharks and rays have been found in some archaeological sites, but because of their small size, special techniques (sieving and floating) are required to find them.

See   Dermal denticles
Fig: 8
Table   4

272. Platybasic skull

Fr: crâne platybasique
Ger: plattbasischer Schädel
Lat:
Rus: платибазальный череп
Sp: cráneo platibásico

The platybasic skull is characterized by having the trabeculae widely separated, as in Cypriniformes. This condition seems to be a primitive one, since it is found in primitive fishes (*Amia, Polypterus,* and *Acipenser*). The separation of the trabeculae affects the relative size of the eyeballs and brain during development. Van

Wijhe (1922) noted that since this character is concerned with the trabecular distance, the skull should be called platytrabic.

273. Pleurodont teeth

Fr: dents pleurodontes
Ger: pleurodonte Zähne
Lat:
Rus: плевродо́нтные зу́бы (sing: плевродо́нтный зуб)
Sp: dientes peurodontos

Teeth implanted in the sides of a bone, as in *Balistes* and *Scarus.*

274. Pleurosphenoid

Fr: pleurosphénoïd
Ger: Pleurosphenoid
Lat: os pleurosphenoideum (pl: ossa pleurosphenoidea)
Rus: плевросфенби́д
Sp: pleurosfenoides

This name, which has been applied sometimes to the pterosphenoid, should be discarded, since it has been demonstrated that it is not homologous with the reptilian *pleurosphenoid,* which has priority.

See   Pterosphenoid

275. Polyspondylous vertebra   *

Fr: vertèbre polyspondyle
Ger: polyspondyler Wirbel
Lat: vertebra polyspondyla (pl: vertebræ polyspondylæ)
Rus: полиспонди́льный позвоно́к (pl: п-ные позвонки́)
Sp: vértebra polispondila

Vertebra with several centra: two in the caudal region of *Amia* and as many as five to nine centra per segment in *Chimaera.*

276. Postcleithrum   *

Fr: postcléithrum
Ger: Postcleithrale
Lat: os postcleithrum (pl: ossa postcleithra)
Rus: постклéйтрум; подключи́чная кость (pl: п-ная ко́сти)
Sp: postcleitro

Mesial to the cleithrum, there are up to three dermal bones of variable shape and size, the *postcleithra,* which belong to the skeleton of the secondary pectoral girdle. Their higher number represents a more primitive condition in fishes, from three on each side (Salmonidae) to two in *Perca,* and only one in Gadidae.

Syn: metacleithrum; postclavicle (Parker)
Figs: 1 E, 3, 13, and 25 A
Bibl: Rojo (1986)

277. Postmaxillary process

Fr: processus post-maxillaire
Ger: postmaxillarer Fortsatz
Lat: processus postmaxillaris (pl: processus postmaxillares)
Rus: постмаксиллярный (=заднечелюстнόй) отрόсток
Sp: proceso postmaxilar

Process located in the middle or in the posterior part of the premaxilla. It is directed upward and backward, thereby preventing the lateral dislocation of the premaxilla when the mouth opens.

Fig: 15

278. Postorbitals

Fr: postorbitaires
Ger: Postorbitalia
Lat: ossa postorbitalia (sing: os postorbitale)
Rus: заднеглазничные (= посторбитáльные) кόсти
      [sing: заднеглазничная (= посторбитáльная) кость]
Sp: postorbitales

Name applied to the two or three dermal bones that sometimes form the posterior rim of the orbit. They in fact belong to the infraorbital series and therefore there is no need to assign them any special name. Nevertheless, the uppermost dorsal postorbital bone (IO6) has a special function, that of joining its infraorbital sensory canal to the supraorbital and lateral line sensory canals, and for this reason it has received the special name *dermosphenotic*. They are usually labelled (SO3, SO4, SO5, SO6) because they were considered to belong to the suborbital series. A new interpretation assigns them to the infraorbital series, and for this reason they are abbreviated in this DICTIONARY as IO3, IO4, IO5, IO6.

See   Circumorbitals; Infraorbitals and Dermosphenotic

279. Posttemporal

Fr: posttemporal
Ger: Posttemporale _
Lat: os posttemporale (pl: ossa posttemporalia)
Rus: постемпорáльная (= задневисόчная) кость
Sp: postemporal

Dermal paired bone located in the uppermost position in the secondary pectoral girdle. It is Y-shaped, with the upper prong resting on the epiotic (= epioccipital) or the supraoccipital and with the lower one connected to either the opisthotic bone in primitive fishes, or the intercalar, as in modern fishes. The body of the bone abuts the supracleithrum.
    The cephalic sensory canal runs through this bone, enters the upper part of the supracleithrum and continues along the flanks of the fish, forming there the lateral line canal.
Syn: suprascapula; suprascapular; supracleithrumn I; supraclavicle I.
Figs: 1 E, 3, 13, 25 B, and 39 A

280. Postzygapophyses

Fr: postzygapophyses
Ger: Postzygapophysen (sing: Postzygapophyse)
Lat: postzygapophyses (sing: postzygapophysis)
Rus: задние сочленовные отростки (pl: задний с-ные
       отростки); постзигапофизы (sing: постзигапофиз)
Sp: postzigoapófisis

Paired process found on the posterior part of the vertebral centrum, which articulates with the prezygapophyses of the following vertebra. In some fishes, there are two dorsal and two ventral postzygapophyses.

Fig: 35 C

281. Prearticular

Fr: préarticulaire
Ger: Praearticulare
Lat: os præarticulare (pl: ossa præarticularia)
Rus: предсуставная кость
Sp: prearticular

Paired dermal bone found in *Amia* , among other fishes, covering the mesial side of the mandible. It bears teeth and has a well-developed coronoid process.

282. Preethmoid

Fr: préethmoïde
Ger: Präethmoid;
Lat: os præethmoideum (pl: ossa præethmoidea)
Rus:
Sp: preetmoides

Name used by Swinnerton (1902) to replace the term *septomaxilla,* which was coined by Sagemehl in 1891 and it is still sometimes used when referring to *Amia.* This change was proposed when it was recognized that the septomaxilla was not homologous with the bone of the same name in tetrapods.

See   Septomaxilla

283. Prefrontal

Fr: préfrontal
Ger: Praefrontale
Lat: os præfrontale (pl: ossa præfrontalia)
Rus: предлобная кость (pl: предлобные кости)
Sp: prefrontal

Paired bone of dermal origin belonging to the circumorbital series. Sometimes the lateral ethmoid and the prefrontal fuse together forming a bone of mixed origin.

See: Lateral ethmoid

284. Premaxilla   *

Fr: prémaxillaire
Ger: Praemaxilla
Lat: os præmaxilla (pl: ossa præmaxillæ)
Rus: предчёлюстная (=премаксиллярная) кость
[pl: предчёлюстные (=премаксиллярные) кости]
Sp: premaxilar; premaxila

The premaxilla or premaxillary is a paired, dermal bone found at the anterior part of the upper jaw. In most fishes, both premaxillae join anteriorly at the maxillar symphysis. In some cases, the premaxillae are ankylosed, thus forming a single bone, as in Diodontidae. In other cases they may overlap or be somewhat separated. The lower border of the premaxilla has teeth of different types according to the fish diet, although some taxa, such as Cyprinidae and Argentinidae, lack them.

The premaxilla is lacking in chondrosteans. The holostean and teleostean premaxillae both have an ascending process at their anterior end, which according to Patterson (1973), are not homologous. In holosteans (*Lepisosteus* and *Amia*), the premaxilla has two ossification centers, and hence is considered a double bone (See Premaxillo-Nasal).

There are several lines in the evolution of the premaxilla of teleosts. In Clupeomorpha and Elopomorpha, the premaxilla is large, movable, and tooth-bearing, forming with the maxillaries the fish gape. The teleost premaxilla have from one to three processes, named (from front to back) the *ascending,* the *articular,* and the *postmaxillary process*. In some cases, there is also present a posterior extension called *caudal process*.

In older literature, this bone has also been called *intermaxillary, surmaxillary,* or *bimaxillary*.

Syn: premaxillary
Figs: 1 B, 3, 13, 14 B, and 15
Bibl: Rojo (1986)

285. Premaxillo-ethmo-vomer

Fr: prémaxillo-ethmo-vomer
Ger: Praemaxillo-etmo-vomer
Lat:
Rus:
Sp: premaxilo-etmo-vómer

In Anguilliformes, the premaxilla is fused to the vomer, the ethmoid, and the lateral ethmoid forming a complex bone, called *premaxillo-ethmo-vomer*.

286. Prenasal

Fr: prénasal
Ger: Praenasale
Lat: os prænasale (pl:ossa prænasalia)
Rus: предносовая кость (pl: предносовые кости)
Sp: prenasal

The anteriormost dermal bone bearing part of the sensory canal in *Lepisosteus*.

Syn: rostral (Aumonier,1942)

287. Preopercle *

Fr: préopercule; préoperculaire
Ger: Praeoperculum; Praeoperculare
Lat: os præoperculum (pl: ossa præopercula)
Rus: преоперкулярная (= предкрышечная) кость
        [pl: преоперкулярные (= предкрышечные) кости];
        предкрышка (pl: предкрышки)
Sp: preopérculo; preopercular

In spite of its name, the preopercle does not belong to the opercular series. It is a paired bone of dermal origin that contains the preopercular branch of the mandibular sensory canal.

In teleosts, the preopercle is located at the back of the suspensorium to which it belongs, at least functionally, since it acts as a reinforcement, preventing it from sliding outward. The anterior border of the preopercle articulates with the symplectic, its laminar part partially covers the three opercular bones. It sometimes bears finely serrated teeth or stout spines on its posterior border.

In *Acipenser,* the preopercle is represented by five to seven ossicles that bear the preopercular canal, albeit the latter is independent of the rest of the system. Parker and Regan call this bone *interopercular* in *Lepisosteus.*

Figs: 1 C, 3, 13, 14 B, and 15.
Bibl: Rojo (1986)

288. Preorbital

Fr: préorbitaire
Ger: Praeorbitale
Lat: os præorbitale (pl:ossa præorbitalia)
Rus: предглазничная кость
Sp: preorbitario

Name given to the first bone of the infraorbital series, equivalent to the lacrymal. Special mention should be made of the preorbital of *Lepisosteus* in which this bone has split into 6 to 8 ossicles, which are joined later by a series of teeth, according to Hammarberg (1937). This assemblage forms the border of the upper jaw. Owing to its position, this series of ossicles has also been considered a maxillary bone divided into several units.

Syn: lacrymal

289. Preural vertebrae

Fr: vertèbres préurales
Ger: praeurale Wirbel
Lat: vertebrae præurales (sing: vertebra præuralis)
Rus: преуральные позвонки (sing: преуральный позвонок)
Sp: vértebras preurales

According to Nybelin (1963), these are the vertebrae that precede the ural vertebrae and consequently lack hypurals.

The vertebra preceding the first ural (U_1) is called first preural vertebra (PU_1); the remaining (PU_2, PU_3, etc.) are counted in the direction of the head.

Fig: 37

290. Prevomer   *

Fr: prévomer
Ger: Praevomer
Lat: os prævomere
Rus: предсошникóвая кость; предсошнйк
Sp: prevómer

Name proposed to replace the most common name *vomer,* since according to Gregory (1933) and other anatomists, this bone is not equivalent to the mammalian vomer.   Although *prevomer* is being used more and more frequently, many ichthyologists still use the old term vomer.

See   vomer
Figs: 14 B and 19

291. Prezygapophyses

Fr: prézygapophyses
Ger: Praezygapophysen (sing: Praezygapophyse)
Lat: præzygapophyses (sing: præzygapophysis)
Rus: передние сочленóвные отрóстки [pl: п-ние с-ные
       отрóстки; презигапóфизы (sing: презигапóфиз)
Sp: prezigoapófisis (sing: prezigoapófisis)

The vertebrae of the teleost fishes often have one or two pairs of anterior processes, the *prezygapophyses,* which allow each vertebra to articulate with the preceding one.   They are arranged in pairs: the dorsal prezygapophyses, above the centrum, and below it, the ventral prezygapophyses.

Fig: 35 C

292. Priapium

Fr: priape
Ger: männliches Begattunsorgan
Lat: ossa priapia
Rus: приáпия кóсти
Sp: priapo

Name proposed by Regan (1916) to designate the mating organ of the oviparous family Phallostethidae.   The paired fins contribute to the complexity of the skeleton of the priapal apparatus. The skeletal elements that form the priapium are very numerous and variable in each species of this family.
In general, they can be grouped into three functional units:
a)   the supporting elements, with the cleithrum, the pulvinar, the priapal ribs, the antepleural cartilage and the axial bone;
b)   the claspers, consisting of the ctenactinium, the toxactinium, the infrasulcar, and the uncus; and
c)   the papillary unit composed of the penial, the basipenial, the papillary, the prepapillary, and the cristate bones.

293. Process

Fr: processus
Ger: Fortsatz (pl: Fortsätze)
Lat: processus (pl: processus)
Rus: отро́сток (пл: отро́стки); прида́ток; вы́рост
Sp: proceso

A process is any prominence or projection that protrudes from the body or mass of a bone. Processes receive different names, according to their size and shape, such as head, condyle, apophysis, tuberosity, etc. (See these terms).

Figs: 15, 16, 17, 18, 19, 20, 24, 29, 30, and 36

294. Proethmoid

Fr: proethmoïde
Ger: Proethmoid
Lat: os proethmoideum (pl: ossa proethmoidea)
Rus: проэтмо́ид
Sp: proetmoides

Starks (1926) assigned the name *proethmoid* to the paired bone of dermal origin located in the antero-dorsal part of the ethmoid region. The proethmoid is sometimes confused with the endochondral *preethmoid*.

295. Prootic   *

Fr: prootique
Ger: Prooticum
Lat: os prooticum (pl: ossa prootica)
Rus: пере́днеу́шная кость
Sp: proótico

Name proposed by Huxley (1858) for the anterior ossification of the otic capsule of actinopterygians. The prootic is the only bone of the otic capsule found in *Amia*

Fig: 2

296. Propterygium

Fr: proptérygium
Ger: Propterygium
Lat: propterygium (pl: propterygia)
Rus: проптери́гий
Sp: propterigio

The outer or anteriormost basal cartilage of the paired fins in cartilaginous fishes. It is often the shortest of the three basal cartilages. In rays and skates, the pectoral propterygium elongates anteriorly to support the numerous radial cartilages of the pectoral fins, which reach extraordinary proportions in these fishes. In some cases, as in the shark *Heterodontus*, the propterygium is wanting.

297. Pterosphenoid *

Fr: ptérosphénoïde
Ger: Pterosphenoid
Lat: os pterosphenoideum (pl: ossa pterosphenoidea)
Rus: крылоклиновидная кость (: крылоклиновидные кости)
Sp: pteroesfenoides

Paired bone of endochondral origin located in the posterior part of the orbitosphenoid region. The term pterosphenoid was proposed by Goodrich (1930) to replace the old name, *alisphenoid*, since this fish bone is not homologous with the mammalian alisphenoid. For a similar reason, the name *pleurosphenoid* should be discarded, since the bone so named in reptiles is not homologous with the pterosphenoid of fishes.

Syn: alisphenoid (Allis, 1903; Gregory, 1933); pleurosphenoid (Beer, 1937).
See   Alisphenoid
Figs: 2, 13, and 14 A

298. Pterotic *

Fr: ptérotique
Ger: Pteroticum
Lat: os pteroticum (pl: pterotica)
Rus: крылоушная кость (pl: крылоушние кости)
Sp: pterótico

1. A paired dermal bone with a sensory canal that occupies the dorsal side of the otic capsule in *Amia*. Holmgren and Stensiö (1936) named it *intertemporo-supratemporo-parietal,* because it is formed by these three bones.
2. According to Parker (1874), the pterotic is a bone of mixed origin that covers the ampulla of the horizontal semicircular canal in actinopterygians. It is formed by an endochondral element, the *autopterotic*, and a dermal unit, with or without sensory canals. Its outer surface has a depression, the *hyomandibular fossa*, where the hyomandibular articulates.

The name *squamous* or *squamosal,* also applied sometimes to this bone, should be reserved for tetrapods.

Figs: 2, 13, and 14 A

299. Pterotic spine

Fr: épine ptérotique
Ger: Pteroticum-Stachel
Lat: spina pterotica (pl· spinæ pteroticæ)
Rus: крылоушный шип
Sp: espina pterótica

A pointed process of the pterotic bone.

300. Pterygiophores *

Fr: ptérygophores
Ger: Pterygophoren (sing: Pterygophore); Radien

Lat: pterygiophori (sing: pterygiophorus)
Rus: птеригофо́ры (sing: пте́рыгофо́р)
Sp: pterigóforos

Pterygiophores are the bony elements that support the radii of the dorsal and anal fins. Very often a pterygiophore appears in embryogeny formed by three independent cartilages that ossify into one, two, or three units, named, proximal or *axonost*, median or *mesonost*, and distal, *epibaseost* or *baseost,* the latter being the closest to the fin. When there are only two pieces, they are known as proximal (axonost) and distal (baseost). When the proximal element is very long, owing to an excessive growth or fusion with other elements, it is known simply as *basal*. In *Hippocampus,* the basal elements fuse with the vertebrae.

The pterygiophore of the dorsal fins or its proximal element is specifically called an *interneural* , while its equivalent for the anal fin is an *interhemal*. The term *actinophore,* proposed by Cope (1890) to refer to the ensemble of the pterygiophore with its corresponding fin ray, has not been widely accepted.

In chondrichthyans and in primitive actinopterygians, the pterygiophores consist of three elements, which also appear in the embryos of most Teleosts. In the most advanced Teleosts, there may be either two (*Esox*) or only one (*Merluccius*).

Pterygiophores have a double function, since they act both as supporters of the fin rays and as an attachment for the erector and depressor muscles.

The vestigial pterygiophores present in some fishes in front of the dorsal fin, receive the special name of *supraneurals*. The last pterygiophore of the dorsal fin is called *stay*.

Figs: 13 and 41 B

301. Pterygoid

Fr: ptérygoïde
Ger: Pterygoid
Lat: os pterygoideum (pl: ossa pterygoidea)
Rus: крылови́дная кость (pl: крылови́дные ко́сти)
Sp: pterigoides

Paired bone of dermal origin formed in the central part of the palatoquadrate (pterygoquadrate) cartilage, between the palatine and the quadrate. The pterygoid forms part of the anterior branch of the suspensorium.

Syn: ectopterygoid

302. Quadrate   *

Fr: carré
Ger: Quadratum
Lat: os quadratum (pl: ossa quadrata)
Rus: квадра́тная кость (pl: квадра́тные ко́сти)
Sp: cuadrado

Paired bone of endochondral origin formed from the posterior part of the palatoquadrate cartilage. In most teleosts, it has a triangular shape, with the lower vertex articulating with the angular bone of the lower mandible. Anteriorly, it joins with the ectopterygoid and dorsally, with the metapterygoid, the mandibular and the symplectic. It acts as a pivot for the suspensorium.

The quadrate has a posterior process forming a narrow angle with the body of the bone in many actinopterygians. This process, according to Patterson (1973), could represent the quadratojugal bone of primitive actinopterygians.

During the evolution of vertebrates, the quadrate was incorporated into the middle ear of mammals as a small bone, the *anvil* or *incus*, as was demonstrated by C. Reichert in the last century.

Figs: 1 D, 3, 13, 25 A, and 29
Bibl: Edgeworth (1923); Rojo (1986)

303. Quadratojugal

Fr: quadrato-jugal
Ger: Quadratojugale
Lat: os quadrato-jugale (pl: ossa quadrato-jugalia)
Rus: квадра́тно-скулова́я кость
Sp: cuadratoyugal

Name given by Hammarberg (1937) to the paired, dermal bone located behind the quadrate. Because it is placed in front of the preopercular, it has also been called *interopercle* or *preopercle*.

It is most likely that the presence of the quadratojugal is a primitive feature in actinopterygians. A quadratojugal is present in *Lepisosteus* and sturgeons, but is missing in *Amia*.

Although it ossifies independently in many teleosts (*Salmo, Syngnathus*) it later joins the quadrate as a process, called the *quadratojugal process* by Holmgren and Stensiö (1936).

304. Quadratojugal process

Fr: processus quadrate-jugal
Ger: Quadratojugalfortsatz
Lat: processus quadratojugalis (pl: processus quadratojugales)
Rus: квадра́тно-скулово́й отро́сток
Sp: proceso cuadratoyugal

Bony process in the lower part of the quadrate bone. The symplectic bone fits into the space left between this process and the body of the quadrate bone.

See   Quadratojugal
Fig: 29

305. Radial formula

Fr: formule radiaire
Ger: Radialformel; Flossenstrahlformel
Lat:
Rus: лучева́я фо́рмула
Sp: fórmula radial

Among the meristic characters of fishes most often used in systematic and comparative studies special mention should be made of fin rays. The number of rays for each fin is determined firstly, by genetic factors that operate at the species and population level, and secondly, by environmental conditions that influence its phenotypic expression.

In order to express, in a simple and graphic way, the number of fin rays of a species, a so-called *radial formula* has been devised. In this aritmethical expression, the minimum and maximum number of rays for each fin in the sample studied is indicated. When a single specimen is used, the actual number of rays of each fin is offered. Evidently, in this case the information provided has less informative value.

The initial letter of the fin name accompanied by a subscript in the case of two fins having identical initial, is used in the formula, as follows: D_1, D_2, D_3, P_1, P_2, A_1, A_2, and C, representing respectively, the first, second and third dorsal, the pectoral, the pelvic, the first and second anal, and the caudal.

Spiny rays, if present, are indicated with Roman numerals, followed by the Arabic numbers for the soft rays. The radial formula for carp according to Mansuetti and Hardy (1967) is then as follows

Fin	Rays
D	III-IV; 14-19
P_1	13-16
P_2	I; 8
A	II-III; 5 - 7
C	19 20

Pectoral fins have more rays than their corresponding pelvic fins which, also very often, have a fixed number of rays. The last two rays of the dorsal and anal fins are counted separately when each one is supported by a single pterygiophore. If both share a common support, they are considered as one (Weitzman, 1962).

306. Radials

Fr: radiaux (sing: radial)
Ger: Radialia
Lat: ossa radialia (sing: os radiale)
Rus: радиалии (sing: радиалия)
Sp: radiales

1. Name used by Goodrich (1930) to designate the supporting endochondral ossicles of the fin rays in fishes. In modern nomenclature, this name is applied exclusively to the bones which support the paired fin rays. Those supporting the rays of the dorsal and anal fins are commonly called *pterygiophores* (see this term). There is a special terminology, for the bones supporting the caudal fin rays. (See Caudal skeleton).

The radials are skeletal elements whose number is related to the number of fin rays, being numerous in the paired fins of primitive actinopterygians (*Polyodon,* 13; *Huso, 10; Acipenser, 9*) and few in modern ones (*Merluccius,* 4; *Lophius, 2*). In the pelvic fins, their number is reduced (*Amia, 4*). In many teleosts the radials disappear in adult life.

2. This term is also applied to the cartilaginous elements supporting the radii of the fins in cartilaginous fishes.

Syn: actinosts
Figs: 1 E, 13, 25 B, and 38 D

307. Rays *

Fr: rayons
Ger: Strahlen (sing: Strahl)
Lat: radii (sing: radius)
Rus: лучи́ (sing: луч)
Sp: radios

Fish fins are supported by rays, rod-like structures arranged serially and provided with a set of thin muscles that allow free movement. Their origin is always dermal, whence the name *dermatotrichs,* by which all types of fin rays are known. Goodrich (1904) classified fin rays into four types according to their nature and structure.

1. The *ceratotrichs,* of cartilaginous fishes (elasmobranchs and holocephalans), have their fins reinforced by cylindrical flexible and non-segmented rays of a fibrous nature.
2. The *actinotrichs* are very thin and made up of a keratinous substance, the *elastoidine*, which makes them similar to the ceratotrichs of the cartilaginous fishes. They appear in the fins in the early stages of teleost embryos and disappear later, remaining only as the support of the finlets of Scombridae, Argentinidae, etc.
3. During the later larval stages of teleosts, the actinotrichs are replaced by bony rays, called *lepidotrichs*. These are the definitive and phylogenetically more recent rays in bony fishes. They are considered modified scales, whose function is to stiffen the delicate fin membranes.
Depending on their structure, the lepidotrichs can be subdivided as follows:
    a)   *spiny rays,* present in the anterior part of the dorsal and anal fins of acanthopterygians. They are always bilateral structures made up of two lateral components paired in the midline, unsegmented, hard, and pointed. They are known also as spines (see this term); and
    b)   *flexible soft rays* , found in malacopterygians, and also in the pectoral and pelvic fins, and the posterior sections of the dorsal and anal fins of acanthopterygians. They are also bilateral, and can be either *simple*, *segmented,* or *branched* at their distal end.

In the caudal fin there are several types of rays classified according to their position and size: 1) the *procurrent rays,* small, unsegmented, and simple, located on the dorsal and ventral edges of the fin; and 2) the *principal* or *main* rays, divided into, a) long, segmented and simple; and b) segmented and ramified occupying the central section of the tail. The upper and lower lobes have sometimes different number of rays, but always both numbers represent a fixed morphological character.

4. *Camptotrichs*. These fin rays, whose structure and nature is midway between the ceratotrichs of cartilaginous fishes and the lepidotrichs of the bony fishes, are present in lung fishes.

The number of rays is an important meristic character, which is used to separate sibling species and even populations of fishes of the same species. Rays have also been used extensively and successfully to estimate the age of some fishes of the following families: Acipenseridae, Catostomidae, Scombridae, and Lophiidae. For an accurate reading of age marks, the rays should be cut a few millimetres from their base.

See   Radial formula
Figs: 6, 13, and 37
Bibl: Arita (1971); Gulland (1958); Rojo and Capezzani (1971); Seymour (1959)

308. Retroarticular *

Fr: rétro-articulaire
Ger: Retroarticulare
Lat: os retroarticulare (pl: ossa retroarticularia)
Rus:
Sp: retroarticular

Small endochondral bone of triangular shape found at the posterior angle of the mandible of *Amia*. The retroarticular of teleost fishes is of endochondral, dermal or even mixed origin. This bone is attached to the angular bone, but does not form part of the mandibular articulation. It was originally called *angular* because of its position at the angle of the mandible, but Böker (1913) rejected this last name and replaced it with the new term *retroarticular*.

Syn: angular (Ridewood, 1904; Gregory, 1933; Berg, 1940); Bridge's ossicle *a* ; lower articular; angulo-retroarticular.
Figs: 1 B, 3, 13, 14 B, and 18

309. Rhomboid scales

Fr: écailles rhomboïdes
Ger: rhomboide Schuppen
Lat:
Rus: четырёхугóльные (= ромбоидáльные) чешуи
        [sing: четырёхугóльная (= ромбоидáльная) чешуя]
Sp: escamas romboideas

Name applied to the ganoid scales on account of their shape.

See   Ganoid scales
Syn: rhombic scales; ganoid scales
Fig: 8 B

310. Ribs   *

Fr: côtes
Ger: Rippen (sing: Rippe)
Lat: costæ (sing: costa)
Rus: рёбра (sing: ребрó
Sp: costillas

Ribs are bony elements in the shape of thin and long rods, arranged serially along the vertebral column and associated with the apophyses of the vertebrae. Their distal end is always free since there is no sternum in fishes.

Depending on their position, ribs can be classified as dorsal and ventral. The *dorsal, epipleural* or *true* ribs are formed at the intersections of the vertical myosepta with the horizontal septum, although this position is not always constant. Chondrosteans and teleosts have rudimentary dorsal ribs.

The *ventral* or *pleural* ribs are formed on the body wall at the level of the vertical myosepta and usually articulate with the parapophyses of the vertebrae. They grow outward and downward, protecting the visceral organs.

The major change that took place during the evolution of the ribs is the loss of the ventral (pleural) ribs during the transition from aquatic to terrestrial life.

Fig: 13
Bibl: Emelianov (1973); Kamel (1952)

311. Rostral

Fr: rostral
Ger: Rostrale
Lat: os rostrale (pl: ossa rostralia)
Rus: межчелюстна́я кость (pl: межчелюстны́е ко́сти)
Sp: rostral

Paired bone of dermal origin bearing a sensory canal. The rostral is located in the anterodorsal part of the ethmoidal region in *Amia* and in primitive teleosts, such as *Elops*. This term was first proposed by Sagemehl in 1874 for the bone now called *ethmoid*.

Syn: ethmoid

312. Rostral teeth

Fr: dents rostrales
Ger: rostrale Zähne
Lat: dentes rostrales
Rus: межчелюстны́е зу́бы (sing: межчелюстны́й зуб)
Sp: dientes rostrales

Teeth implanted on the sides of the rostrum of sawsharks (Pristiophoridae) and sawfishes (Pristidae), groups which belong to the Selachii and Batoidei, respectively. Rostral teeth are made of orthodentine and vitrodentine with the pulp cavity filled with osteodentine, but lack the basal plate typical of other elasmobranch teeth. They are characterized by being true thecodont teeth, since they are well implanted in the sockets or *alveoli*. They are morphologically and histologically more similar to mammalian teeth than to elasmobranch teeth.

See   Thecodont teeth

313. Rostrum

Fr: rostre
Ger: Rostrum
Lat: rostrum (pl: rostra)
Rus: ро́струм; ры́ло
Sp: rostro

1. The anterior part of the snout when it is prolonged forward, as in sharks, rays, skates, sturgeons and paddlefishes. In these cases the mouth occupies a ventral position.

2. The bony shelf overhanging the cavity where the esca of the ceratiid fishes is lodged. (See Illicium).

3. The longer lip of the ostium of the otolith sagitta.

Fig: 43

314. Saccule

Fr: saccule
Ger: Säckchen; Sakkulus (= Sacculus)
Lat: sacculus (pl: sacculi)
Rus: мешóчек
Sp: sáculo

The saccule is the largest chamber of the membraneous labyrinth. It is connected to the upper part of the labyrinth (utricle and semicircular canals) in lampreys, cartilaginous fishes, and primitive bony fishes, but in more advanced teleosts, the saccule is joined to the upper part of the labyrinth by a narrow connexion (cyprinoid type). In exceptional cases (Aplodinotus), there is complete separation between both parts. The saccule encloses the saccular macula, on which rests the sacculolith or sagitta.

See   membraneous labyrinth
Fig: 42

315. Sagitta   *

Fr: sagitta
Ger: Sagitta
Lat: sagitta (pl: sagittæ)
Rus:
Sp: sagita

The sagitta, also called *saculolith*, is found in the saccule of rayfinned fishes. Its name refers to its shape, which vaguely resembles that of an arrowhead. It is usually the largest of the three otoliths (except in Cyprinidae and Siluridae), and is therefore most frequently used in ageing fishes.

The morphology of the sagitta, which has been intensively studied, is useful for recognizing fossil species; for taxonomic identification of modern fishes; for diet studies of marine animals (fish, birds, mammals); and for the analysis of fish remains in archaeological settlements.

In most cases, the sagitta has an anterior pointed end, while its posterior end is notched, rounded, or straight. Its mesial side has a groove, the *sulcus acusticus,* which extends in most cases from end to end (*Oncorhynchus*), dividing the sagitta into dorsal and ventral fields. The sulcus expands into a wide entrance, the *ostium,* framed by two pointed ends, the most advanced being the *rostrum,* and opposite it, the *antirostrum,* which is shorter. The posterior end or *cauda* sometimes ends short of the posterior rim. Often, two prominent points are formed at the end of the cauda: the *pararostrum,* above, and the *postrostrum,* below. The posterior branch of the acoustic nerve runs along the sulcus, which is delimited by the *superior* and *inferior crests.* A narrowing of the sulcus, called *collum* , marks the position of the initial nucleus or focus of the formation of the sagitta. This is precisely the place where a cut is made in the sagitta for age reading. A cut that does not pass through the nucleus is liable to miss one or more growth rings.

The sagitta is held in the saccule in a vertical position, with the rostrum pointing forward and its concave side arching outwards. The upper rim is shorter than the lower or ventral rim. Frequently, the rims have arched bulges, called *domes,* or a continuous scalloped outline.

There is no significant difference either for the weight or the length between the right and left otoliths in either sex of many species (Rojo 1977). This fact allows us to use either sagitta to calculate the live length or weight of the fish with a high degree of confidence, unless it has been eroded badly since the time of its deposition on the ground.

See   Otoliths and Statoliths
Syn: Sacculolith; sacculith
Figs: 42, 43, and 44

316. Scales   *

Fr: écailles
Ger: Schuppen
Lat: squamæ (sing: squama)
Rus: чешуи (sing: чешуя)
Sp: escamas

The scales of modern fishes are skeletal elements derived from the dermal exoskeleton of ostracoderms and placoderms. In placoderms, abundant during the Devonian, the armour of large plates was fragmented into smaller units which allowed for greater flexibility of movements and a more active life in the water, thus freeing them from their demersal habitat. From this dermal skeleton derived three structurally different elements: scales, teeth and certain bones of the dermocranium and of the secondary pectoral girdle. The main lines of this evolutionary process are indicated in summarized form in table 4.

Although not indicated on the table, it must be kept in mind, that in any group there may be a loss of one or another of the above mentioned elements. For example, there are scaleless fishes (Siluridae); toothless fishes, such as some Clupeidae; and fishes, like the sharks, that completely lack a dermocranium.

A degenerative process has taken place in the evolution of the scales, starting with the thick and large plates of the armoured fishes and ending in the thin and flexible scales of actinopterygians. Even within this last group there is a strong contrast between the relatively thick and hard scales of chondrosteans and holosteans, on the one hand, and the teleostean scales, on the other. Among teleosteans there are species with vestigial scales and others which lack them altogether.

Modern fish scales can be grouped into four types following a double criterion, evolutionary and structural, the latter based on histological study and chemical analysis. These four groups are: placoid, cosmoid, ganoid, and leptoid. Since cosmoid scales are only present in fossil fishes and never associated with human remains, they are excluded here. The three remaining types are:

1. placoid scales of elasmobranchs, microscopic and difficult to recover in archaeological sites, except by special methods (fine screening and water suspension);

2. ganoid scales of primitive actinopterygians: subdivided into
   a) paleoniscoid scales of brachiopterygians; and
   b) lepisosteoid scales of *Lepisosteus* ;

3. leptoid or elasmoid scales of teleosteans: subdivided into
   a) cycloid scales of malacopterygians; and
   b) ctenoid scales of some acanthopterygians.

In systematic studies, it is important to know the number of scales running along the lateral line and those above and below it. It is customary to indicate the number of scales along the lateral line by counting the scales with pores or the scales up to the end of the caudal peduncle. The count of scales above the lateral line is taken starting at the origin of the dorsal fin, and following downward and backward, down to the lateral line, but excluding from the count the lateral line scale. The count of scales below the lateral line is done upward and forward, starting at the base of the anal fin, again excluding the lateral line scale. (Fig: 6). The values obtained are presented as a mixed fraction, for example,

$$44 \; \frac{16}{22}$$

where the whole number represents the number of lateral line scales; the numerator, those above the lateral line; and the denominator, the scales below that line. Since there is an interspecific as well as an intraspecific variation in the number of scales in any one species, it is customary to give the maximum, the minimum and the mean value for each one of these three values in the sample studied.

Biological and archaeological applications of scale study

The first use of scales to determine age was on carp (*Cyprinus carpio*) in 1898. Scales have certain features which have been used to estimate not only age but also growth and other biological circumstances of the life of fishes, such as spawning activity, migrations, diseases, environmental conditions related to temperature, lack of food, etc.

1. Age estimation

Ganoid, cycloid, and ctenoid scales grow by deposition of material around an initial nucleus (*focus* or *center*) located in the center or in one of the foci of the oval or elipse. The outer surface can be divided into two main fields, which are sometimes limited by radii. These fields are: the *anterior* or *oral*, which is covered by the previous scale and the *posterior* or *aboral* field, which is free and which covers the anterior field of the following scale, and is itself covered by the epidermis (Fig. 10).

The outer surface of the scales shows concentric growth lines, called *circuli*, (sing: *circulus*) in the form of continuous or discontinuous ridges. In the latter case each section of the concentric lines is called *sclerite*. The separation between the growth lines is greater during the summer than during the winter. The result is an alternating pattern of thick and thin bands. The line or group of lines formed during the period of physiological slowdown, which usually corresponds to the winter season, is called an *annulus* (pl: *annuli*). A simple count of the annuli gives an estimation of the fish age, which is always given in whole numbers. If there is a marginal zone representing partial growth during the year when the fish was captured, the number representing the age is followed by a plus sign (+).

Although the wide band is referred to as "summer" growth, this period can extend from early spring until late fall depending on the species, the geographical area, water temperature, abundance of food and individual condition of the fish.

Even though it is accepted that the concentric structures of the scale surface represent units of time, it is advisable to establish, by reliable experiments or observations, the interrelationship of annuli and years. Regier (1962) and Cable (1956) have validated the use of scales for ageing in bluegill (*Lepomis macrochirus*) and lake trout (*Cristivomer namaycush*), respectively. Nichy (1974) offers a list of fishes for which the use of scales for ageing purposes has been reliably validated.

In some cases, counting annuli is a simple operation, but it usually requires a good knowledge of the biology of the fish. It is often difficult for a novice scale reader, whether a biologist or not, to recognize yearly patterns of scale growth, due to the wide variation in the growth pattern of individual fishes and of fish from different water masses.

The most important commercial fish families from which scales are taken for age studies are the following: Elopidae, Salmonidae, Esocidae, Cyprinidae,

Catostomidae, Cyprinodontidae, Percichthyidae, Serranidae, Centrarchidae, Sparidae, Sciaenidae, Cichlidae, Gadidae, Bothidae, Pleuronectidae, and Mugililidae.

Besides the scales, fishes have other structures whose growth can be detected by parallel lines or concentric annuli, thus offering the possibility of age estimation. These structures are: otoliths, vertebrae, spiny fin rays, scutes, and certain bones (opercle, cleithrum, etc.). (See the entries corresponding to these structures).

2. Growth studies

Since scales grow proportionally to the length of the fish, it is possible by means of simple calculations to estimate the fish length corresponding to each year prior to its capture. The proportionality between fish length and scale size can be expressed by the following series of fractions:

$$\frac{L_t}{S_t} = \frac{L_1}{S_1} = \frac{L_2}{S_2} = \frac{L_3}{S_3} = \ldots \ldots \frac{L_n}{S_n}$$

where L_t is the fish size at the time of capture; S_t, the radius or diameter of the scale at that moment; $L_1, L_2, L_3, \ldots \ldots L_n$, the consecutive lengths of the fish at ages $1, 2, 3$ .... n years; and $S_1, S_2, S_3, \ldots S_n$, the scale radius or diameter for those years represented in the scales by the $1st, 2nd, 3rd,\ldots n^{th}$ annulus. Thus, in order to calculate, for example, the size of the fish in its third year, the following formula can be used:

$$L_3 = \frac{L_t \ \times \ S_3}{S_t}$$

For consistency and accuracy of the results, the scale radius or its diameter should always be taken in the same direction in all scales.

The above formula indicates that the relationship between fish length (L) and scale size (S) is of the type

$$L = a \ \times \ S$$

a relationship first proposed by Dah (1909) and Lea (1910).

Lee (1920) noticed that using this formula, the fish length calculated for the early years of a fish is smaller when using the length of old specimens than when using younger specimens. She attributed this phenomenon to a shift in the growth rate. In order to correct this apparent error, Lee proposed the formula

$$L = a + b \times S$$

This new proportionality can be expressed with this new series of fractions:

$$\frac{L_t - a}{S_t} = \frac{L_1 - a}{S_1} = \frac{L_2 - a}{S_2} = \frac{L_3 - a}{S_3} = \ldots \ldots \frac{L_n - a}{S_n}$$

where a, the ordinate intercept, represents in some cases the fish length at the moment of the appearance of the first scales.

Therefore, the length of the fish at age three, will be

$$L_3 = \frac{(L_t - a) \times S_3}{S_3} + a$$

Lee's phenomenon, i.e., that smaller estimations of fish length in previous years are found when using older specimens, has been attributed to several causes. Among them are the use of an incorrect formula, lack of randomness in the sample collected, selectivity due to natural and fishing mortalities, etc.

Monastyrsky (1930) described the relationship fish length/scale length with a parabolic equation

$$L = a \times S^n$$

which has been successfully applied in many cases. He demonstrated that the value of the Y intercept (*a*) in Lee's equation does not represent the size of the fish at the moment of scale formation, nor any other morphological value.

Other polynomial equations have been proposed by Sherriff (1922)

$$L = a + bS + cS^2$$

and Carlander (1950)

$$L = a + bS + cS^2 + dS^3$$

but they were not favoured because of their difficult application. For a more detailed study of this matter, see Shuck (1949) and Hile (1970).

Segerstråle (1933) recommends the use of empirical curves obtained from actual measurements of scales of fish of all sizes. These curves should be specific for each population, season, and year since the relation fish length/scale diameter varies according to the environmental factors. From a statistical view point, these differences, even for the same species, can be significant (Warner and Harvey, 1961).

For this reason, when studying archaeological fish remains, it is advisable to use modern data taken from a population as close as possible to the geographical area of the site. Similarly, if the environmental conditions of the old site are known, modern data should be collected from a place of similar ecological conditions.

3. Fish life history studies.

Scales provide a reliable record of many of the contingencies of the life of a fish. In salmon, for example, spawning activity is marked on the scales by a degeneration of the scale edge, caused by the resorption of scale material due to the deep stress required for the spawning migration and the maturation of sexual cells. In the scales, it is also possible to detect the annual migrations from river to sea and vice versa. The growth corresponding to freshwater and marine periods of the life of the fish can also be detected. Other circumstances, such as lack of food, a decrease in temperature during the feeding and breeding seasons, migrations, and diseases, leave a mark in the form of false annuli called *checks*, easily recognized since they are narrow and do not form a complete circle. Food abundance and favorable environmental conditions are detected by a greater width in the summer bands.

4. Squamation

Scales do not appear all at once in the fish body, but begin to form in one or several regions of the body, in a precise pattern which is variable for each species. From this point or points (*focus* or *foci*) new ones are added until the body is completely covered. This process takes some time to complete, in some cases as much as two years. The scales on different regions of the body have varied shapes, structure, and are arranged in slightly different patterns. For this reason, in the study of fish age and growth, it is advisable to use scales from the region where scales first appear and the place where they have regular shape.

Figs: 6, 8, 10, and 11
Table   4
Bibl: Casteel (1973; 1975; 1976); Cockerell (1913); Franklin and Smith (1960); Goodrich (1907); Graham (1929); Hanna (1981); Martin (1982); Phillips (1948); Priegel (1964); Rojo (1983); Rojo y Ramos (1986); Yerkes (1977)

317. Scapula

Fr: scapula
Ger: Schulterblatt; Scapula
Lat: os scapulum (pl: ossa scapula)
Rus: лопáтка; скапуля́рная кость (pl: лопáтки; ск-ные кóсти)
Sp: escápula

Paired endochondral bone belonging to the primary pectoral girdle. It leans anteriorly on the cleithrum and meets ventrally the coracoid. Posteriorly, the scapula supports two or three of the radials of the pectoral fin.
This bone sometimes presents the scapular foramen that allows the passage of nerves and blood vessels. This foramen is sometimes formed by two matching notches present in the scapula and coracoid.

Figs: 1 E, 3, 13, 25 B, and 38 A

318. Scapular cartilage

Fr: cartilage scapulaire
Ger: Scapula-Knorpel
Lat: cartilago scapularis
Rus:
Sp: cartílago escapular

A rod-like cartilage that forms the lateral portion of the coracoscapular bar in the elasmobranch fishes. It articulates ventrally with the coracoid cartilage and dorsally with the suprascapular. On its lateral surface, it bears the glenoid cavity, to which the pectoral fin is attached.

319. Sclerites

Fr:
Ger: Skleriten (sing: Sklerite)
Lat:
Rus: склери́ты (sing: склери́т)
Sp: escleritos

The cycloid and ctenoid scales show in their external surfaces continuous or interrupted concentric ridges or crests in the form of circles or ovals. These ridges represent growth marks. In the case of a discontinuous pattern, the ridges are made of small units called *sclerites*.

320. Sclerotic bones   *

Fr: os sclérotiques
Ger: sclerotische Knochen; Scleroticalia
Lat: ossa scleroticalia (sing: os scleroticale)
Rus: склеротические (= отвердёвшие) кости
Sp: escleróticos

The sclera of the eye sometimes ossifies, developing bones either from its cartilaginous or from its fibrous region. In some cases (*Salmo*) the sclerotic bones occupy an anterior and a posterior position in the eye. In *Gasterosteus*, the positions are dorsal and ventral. In *Xiphias*, the whole sclera ossifies into a bony capsule with two openings, one anterior for the cornea and one posterior, for the passage of the optic nerve.

Fig: 14 B

321. Scutes   *

Fr: boucliers
Ger: Schilde (sing: Schild)
Lat: scuta (sing: scutum)
Rus: щитки (sing: щиток); киловатые чешуи (sing: к-тая чешуя)
Sp: escudos

Hard and thick plates found on the skin of some fishes. The scutes generally have a circular or discoidal shape, with or without spines. They are derived from scales and represent an advanced or specialized characteristic in the evolution of fishes. Sturgeons have five rows of scutes, one dorsal, two lateral and two ventral. In sticklebacks (Gasterosteiformes), the scutes cover the sides of the fish.

Fig: 9
Bibl: Benecke (1986); Brinkhuisen (1986)

322. Semicircular canals

Fr: canaux semi-circulaires
Ger: Bogengänge
Lat: canales semicirculares (sing: canalis semicircularis)
Rus: полукружные каналы (sing: полукружный канал)
Sp: canales semicirculares

All gnathostome fishes have three semicircular canals in the ear capsule: two vertical and one horizontal. Among jawless fishes, hagfishes have only one semicircular canal, while lampreys, have both vertical canals but lack the horizontal canal.
The three canals and the chambers of the otosac below them form what has been called the membraneous labyrinth. The two vertical canals often share a common stalk, the *crus commune*, located in a mesial position. The anterior vertical canal (angled forward and outward) and the posterior (angled backward and outward) are set at an

angle a little larger than 90 degrees, with its vertex towards the median line. The anterior canal of the right side follows the same direction as the posterior vertical of the left side, and vice versa.

At their distal ends, the vertical canals bear an expansion, the *ampulla*. In most fishes, both vertical canals are joined at their bases by the horizontal canal, which has its ampulla in an anterior position. In elasmobranchs, the horizontal canal joins only the anterior vertical canal.

See Membraneous labyrinth
Fig: 42

323. Septomaxilla

Fr: septomaxillaire
Ger: Septomaxilla
Lat: os septomaxillare (pl: ossa septomaxillaria)
Rus: септо-верхнечелюстная кость
Sp: septomaxila

Name given by Sagemehl (1891) to the paired bone of endochondral origin of *Amia* , formed in the anterior part of the ethmoid region, in the so called *cornua trabecularum*. The septomaxilla is located at the base of the olfactory capsules. Swinnerton (1902) changed this name to *preethmoid,* because of the dubious homology of the bone to the septomaxilla of tetrapods.

See Preethmoid

324. Sesamoid bones

Fr: os sésamoïdes
Ger: Sesamknochen
Lat:
Rus: сезамовидные кости (sing: сезамовидная кость)
Sp: huesos sesamoideos

Sesamoid bones are supernumerary bones formed either by fragmentation of a preexisting bone or by the presence of new ossifications with no phylogenetic value. Examples are the sesamoid articular and the sesamoid palatine.

See Coronomeckelian
Table 1

325. Skeletal system

Fr: système squelettique
Ger: Skelettsystem
Lat:
Rus: скелётная система
Sp: sistema esquelético

The skeletal system of fishes comprises a wide variety of tissues and organs of diverse form, consistency, position, and embryonic origin. The common functions of the skeletal tissues are :

a)   to give support to the body;
b)   to provide protection to the most vital organs; and
c)   to give leverage to the striated muscles.

From the structural point of view, the skeleton is made up of the notochord and the connective, cartilaginous, and bony tissues, with a predominance of one or another, depending on the phylogenetic level of the group or on the developmental stage of the fish.

326. Skeleton

Fr: squelette
Ger: Skelett
Lat: skeleton; sceleton (pl: skeletones; sceletones)
Rus: скелéт
Sp: esqueleto

The fish skeleton, considered in a general sense, is comprised of the connective, notochordal, cartilaginous, and bony tissues, with predominance of one or another, depending on the developmental stage of the fish or the phylogenetic level of the group to which it belongs. In a more restricted sense, the skeleton is the assemblage of all cartilaginous and bony elements.

In Cyclostomata and Chondrichthyes, the skeleton remains cartilaginous for the entire life of the fish. In osteichthyans, the embryo starts with a predominance of connective tissue. Later, cartilage predominates during the larval stages, and in the adults, bony tissue is most abundant. The degree of ossification of the skeleton depends on the age of the specimen and its taxonomic level.

Using the gradual ossification of the fish skeleton, as a criterion, actinopterygians were divided into chondrosteans, holosteans, and teleosteans. This classification has been rejected, because of its heavy dependence upon a single character, thereby overlooking many aspects of fish anatomy and biology. Moreover, the separation between these groups is not as clear as was originally supposed, especially when the fossil evidence is considered.

The skeleton can be subdivided into smaller units according to the origin, the structure and the position of the elements comprising them, as is shown in table 5.

Fig: 13

Table 5. Major divisions of the fish skeleton

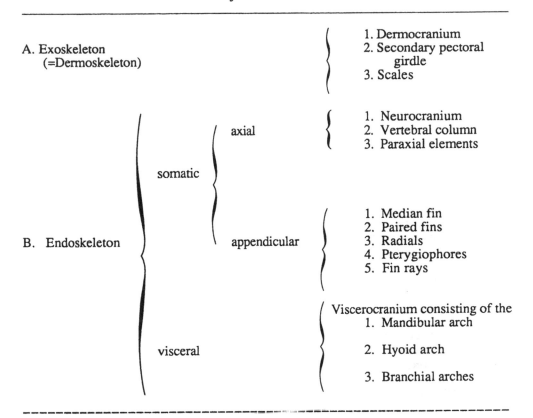

A. Exoskeleton
   (=Dermoskeleton)

1. Dermocranium
2. Secondary pectoral girdle
3. Scales

B. Endoskeleton

somatic

axial
1. Neurocranium
2. Vertebral column
3. Paraxial elements

appendicular
1. Median fin
2. Paired fins
3. Radials
4. Pterygiophores
5. Fin rays

visceral

Viscerocranium consisting of the
1. Mandibular arch
2. Hyoid arch
3. Branchial arches

327. Skull

Fr: crâne
Ger: Schädel
Lat: cranium (pl: crania)
Rus: чёреп ; мозговáя корóбка (pl: мозговы́е корóбки)
Sp: cráneo

The skull is the assemblage of cartilages and bones directly related to the brain, the sense organs of smell, sight, and hearing and finally, to the mouth and gills. This assemblage can also be called *syncranium*.

The skull forms the anterior part of the axial skeleton. Its two main general functions are:

a)   to provide support and shelter to the brain and to the most important sense organs, by means of the *neurocranium* and the *dermocranium*; and

b)   to support the anterior part of the digestive system and the respiratory apparatus by means of the *splanchnocranium* or *viscerocranium*.

The skull of bony fishes ossifies in a greater or lesser degree depending on their phylogenetic level and ontogenetic development. During the early stages of development the skull is completely cartilaginous. As the fish develops, the ossification intensifies, but is never completed because in the adults of many fishes there are extended areas of cartilage. The sum of all cartilaginous elements in the development of the fish head forms the *chondrocranium,* while all bony elements are known by the term *osteocranium.* The former is more extensive in the early stages of development; the latter predominates in the adult stages.

The skull can be divided according to the evolutionary or phylogenetic origin of its elements into the following three units:

a)   *Neurocranium,* formed by the bones directly related to the brain and the sense organ capsules;

b)   *Dermocranium,* comprising the dermal bones that cover and sometimes replace the bones of the other two units; and

c)   *Viscerocranium,* formed by the mandibular, the hyoid and the branchial arches.

See the terms in italics
Syn: cranium
Figs: 2, 3, 4, 14 A and B, and 25 A
Bibl: de Beer (1937); Devillers (1958); Gregory (1933); Harrington (1955); Libois *et al.* (1987); Mujib (1967); Parrington (1967)

328. Sphenotic   *

Fr: sphénotique
Ger: Sphenoticum
Lat: os sphenoticum (pl: ossa sphenotica)
Rus: сфено́тикум
Sp: esfenótico

Parker (1874) gave this name to the bone that separates the orbital from the otic region in the back of the skull. This bone covers the anterior semicircular canal.

See   Autosphenotic
Figs: 2, 13, and 14 A

329. Spines   *

Fr: épines
Ger: Stacheln (sing: Stachel) ; Dornen (sing: Dorn)
Lat: spinæ (sing: spina)
Rus: колю́чки (sing: колю́чка) ; шипы́ (sing: шип)
Sp: espinas

1. Any rigid and sharp process found in the fish body, most commonly related to the bones and scales. The porcupine fish (*Diodon histryx*) provides one of the most striking examples, since spines cover its entire body.

2. This term is also applied to the *ossified, unsegmented and simple* fin rays of the dorsal and anal fins, especially in acanthopterygian fishes, in which they are sometimes connected to poison glands. These rays should not be confused with the *spiny fin rays* , which are formed by two hemitrichs.

The second dorsal fin spine of dogfishes (*Squalus*) and the pectoral spines of Ictaluridae have been used successfully for age determination.

3. This term is also applied to any long and pointed expansion of a bone. In some cases, the number and shape of the spines, as with the preopercular spines of

redfish (*Sebastes*) and sculpins, are used for taxonomic purposes. These spiny processes are sometimes bent at their free end or are provided with hooks.

Figs: 6, 13, and 22
Bibl: Holden (1962); Rosenlund (1986)

330. Splanchnocranium

Fr: splanchnocrâne
Ger: Eingeweideschädel
Lat: splanchnocranium (pl: splanchnocrania)
Rus: спланхнокрániuм
Sp: esplancnocráneo

The splanchnocranium includes all the cartilages and bony elements of the visceral skeleton which have been incorporated to form the skull during the evolutionary transition of the agnathans to the gnathostomes (See table 5).
The first visceral arch added to the skull of placoderms was the mandibular arch that forms the core of the mandibles. In their descendants, the dorsal part (hyomandibular bone) of the hyoid arch (second visceral arch) was also incorporated into the skull, while its ventral section, formed by the hypohyals (dorsohyal and ventrohyal), the ceratohyal (anterohyal) and the epihyal (posterohyal), continues to form part of the visceral skeleton.

Syn: viscerocranium
See Visceral skeleton
Fig: 25 A
Table 5

331. Splenials   *

Fr: spléniaux (sing: splénial)
Ger: Splenialia
Lat: ossa splenialia (sing: os spleniale)
Rus: сплеnáльные (=крышечные) кóсти
Sp: spleniales

These bones, common to fossil osteichthyans, are now restricted to some primitive bony fishes, such as *Amia* . The splenials are dermal bones bearing sensory canals.

Syn: submandibular (Fürbringer); sesamoid angular

332. Squamosal   *

Fr: squamosal
Ger: Squamosum
Lat: os squamosum (pl: ossa squamosa)
Rus: чешýйчатая кость (pl: чешýйчатые кóсти)
Sp: escamoso

Paired bone of endochondral origin, present in fossil crossopterygians and actinopterygians. It is found only in the living crossopterygian (*Latimeria*). It has disappeared in modern actinopterygians or it has been incorporated in the preopercle.

333. Standard length

Fr: longueur standarde
Ger: Standardlänge
Lat:
Rus: стандáртная длинá
Sp: longitud estándar

1. Jordan (1905) defined standard length as the distance in a straight line between the extremity of the snout and the anterior prominence of the urostyle. This procedure excludes from the standard length the last skeletal element of the vertebral column. The measurement so defined is used in studies of systematics in preference to the total length, because it is not affected by the broken or eroded caudal fin rays or other tail anomalies.

2. In works of fishery biology, the standard length must be taken up to the end of the vertebral column, i.e., to the base of the central fin rays.

Since, in practice, many authors have not followed these criteria in their research, it is advisable to describe the method used when taking the standard length. In biological research, the standard length has been taken by different authors from the snout of the fish to the following anatomical landmarks indicated in figure 7:
the beginning of the urostyle bone (AD);
the end of the urostyle (AO); and
the end of the caudal musculature (AE).

Other ways to measure the standard length have also appeared in different biological works, as follows:
to the last scale;
to the last scale with a pore; and
to the beginning of the caudal fin rays.

Fig: 7
Bibl: Royce (1942)

334. Subocular shelf

Fr: console oculaire
Ger:
Lat:
Rus: подглазнúчная полóчка
Sp: repisa ocular

Bony lamina that extends horizontally inwards from the bones of the infraorbital series of certain fishes, such as in Myctophidae.

335. Subopercle   *

Fr: sous-opercule
Ger: Suboperculum
Lat: os suboperculum (ossa subopercula)   os suboperculare (pl: ossa subopercularia)
Rus: субоперкуля́рная (= подкры́шечная) кость; подкры́шка [pl: с-ные (= п-ные) кóсти; подкры́шки]

Sp: subopérculo; subopercular

A paired dermal bone that reinforces the center of the opercular membrane between the opercle and the interopercle. In embryonic stages, it resembles a branchiostegal ray, that because it expands later, is considered to have evolved from the branchiostegal series. Like the remaining bones that make up the opercular membrane, it is flat and thin.

Syn: subopercular
Figs: 1 C, 3, 13, 14 B, and 24

336. Suborbitals

Fr: sous-orbitaires
Ger: Suborbitalia
Lat: ossa suborbitalia (sing: os suborbitale)
Rus: суборбита́льные (= подглазни́чные) ко́сти
Sp: suborbitarios

According to Stensiö (1947), this term should be reserved for the chain of small bones found ventral to the infraorbitals. They are not related to the infraorbital sensory canal. This series is present in the fossil paleoniscoid fishes, while advanced fishes usually lack suborbitals, according to this new interpretation.

See   Infraorbitals

337. Supraangular   *

Fr: supra-angulaire
Ger: Supraangulare
Lat: os supraangulare (pl: ossa supraangularia)
Rus: надуглова́я кость
Sp: supra-angular

Paired bone of membraneous origin present in the posterior section of the inner side of the lower mandible, dorsal to the angular. The supraangular is found in *Amia* and *Lepisosteus*.

338. Supracleithrum   *

Fr: supra-cleithrum
Ger: Supracleithrum
Lat: os supracleithrum (pl: ossa supracleithra)
Rus: надключи́чная (= супраклейтра́льная) кость
Sp: supracleitro

Paired dermal bone forming part of the secondary pectoral girdle. It articulates dorsally with the posttemporal and ventrally with the cleithrum. In the absence of the posttemporal, the supracleithrum serves as the link between the skull and the pectoral girdle.

Chabanaud calls it *hypercleithrum,* using the argument that the *cleithrum* is a Greek term and therefore should be preceded by the Greek prefix *hyper.* This argument can be invalidated by arguing that the term cleithrum is the latinized form of *cleithron,* and consequently should be preceded by the Latin prefix, *supra.*

Figs: 1 E, 3, 13, 25 B, and 39 B

339. Supraethmoid   *

Fr: supra-ethmoïde
Ger: Supraetmoid
Lat: os supraethmoideum (pl: ossa supraethmoidea)
Rus: супраэтмбид
Sp: supraetmoides

Name proposed by de Beer (1937) for the median intramembranous bone resulting from the fusion of two symmetrical laminar bones, which sometimes fuse also with the ethmoid.

Syn: dermethmoid (Gregory, 1933); dermal ethmoid (Harrington, 1955); mesethmoid (Berg, 1940; Chapman, 1941); dermal mesethmoid rostral (Sagemehl, 1885).

340. Supramaxilla   *

Fr: supra-maxillaire
Ger: Supramaxillare
Lat: os supramaxillare (pl: ossa supramaxillaria)
Rus: супрамаксиллярная (=верхнечелюстнāя) кость.
Sp: supramaxila; supramaxilar

Laminar paired bone of dermal origin located dorsal to the posterior part of the maxillary. The supramaxilla is frequently found in malacopterygians but is lost in more advanced teleosts. Salmonidae have one supramaxilla and Sternoptychidae, two.
In certain genera of Clupeidae, a paired, small, usually tooth-bearing bone, called *hypomaxillary,* is located in the gape, behind the premaxillary and below the maxillary (Berry, 1966)

Syn: supramaxillary; surmaxillary; surmaxilla (Ridewood, 1904); jugal (Allis; de Beer; Parker, (1874); malar (Parker, 1874).
Fig: 14 B

341. Supraoccipital   *

Fr: supraoccipital
Ger: Supraoccipitale
Lat: os supraoccipitale (pl: supraoccipitalia)
Rus: верхнезатылочная (= супраокципитāльная) кость.
         [pl: верхнезатылочные (= супраокципитāльные) кбсти]
Sp: supraoccipital

Median bone of mixed origin found in the dorsal part of the occipital region. The supraoccipital forms the upper rim of the foramen magnum. It is absent in chondrosteans and holosteans.
In the remaining bony fishes it develops from the ossification of the otic roof (*tectum synoticum*) and the connective median septum that separates the trunk muscles of the nape. This last membraneous ossification gives origin to the occipital crest, very frequent in teleosts as a point of attachment for the nape muscles.
Figs: 2, 13, and 14 A

342. Supraorbitals *

Fr: supra-orbitaires
Ger: Supraorbitalia
Lat: ossa supraorbitalia (sing: os supraorbitale)
Rus: надглазни́чные ко́сти (sing: надглазни́чная кость)
Sp: supraorbitarios

Small dermal bones that form the superior border of the orbits and which lack sensory canals (Stensiö, 1947). The primitive actinopterygians (Pholidophoridae) had three supraorbitals, while in modern fishes there are at the most two (*Salmo*), although many species have only one. *Amia* and the more advanced teleosts lack them.

See: Circumorbitals
Fig: 14 B

343. Suprapharyngobranchials

Fr: suprapharyngobranchiaux
Ger: Suprapharyngobranchialia
Lat: ossa suprapharyngobranchialia (sing: os suprapharyngobranchiale)
Rus: супрафарингобранхиа́льные ко́сти (sing: с-ная кость)
Sp: suprafaringobranquiales

See   Pharyngobranchials

344. Suprapreopercle *

Fr: supra-préopercule; supra-preoperculaire
Ger: Suprapraeoperculum
Lat: os suprapræoperculum (pl: ossa suprapræopercula)
Rus:
Sp: suprapreopérculo; suprapreopercular

Paired bone of dermal origin that encloses the dorsal section of the hyomandibular canal and is located dorsal to the preopercle in Salmonidae, Anguillidae, or fused to the opercle in Cyprinidae, among other families.

Syn: supratemporal (Parker, 1874); subtemporal (Gregory, 1933); suprapreoperculum (Bruch, 1861; Holmgren and Stensiö, 1936).
Fig: 14 A

345. Suspensorium

Fr: suspenseur
Ger: Kieferstiel; Aufhängeapparat
Lat: suspensorium (pl: suspensoria)
Rus: подве́сок
Sp: suspensorio

The suspensorium is the assemblage of cartilages and bones which link the mandibles of gnathostome fishes to the neurocranium. In bony fishes, it has a "V" shape, with the anterior arm formed, in most cases, by the palatine, endopterygoid and ectopterygoid, and the posterior arm, by the quadrate, symplectic, metapterygoid and hyomandibular. In a stricter sense, the term *suspensorium* refers only to the posterior arm. The pivotal point of both branches lies at the junction of the angular with the quadrate. The suspensorium is braced against a possible outward displacement by the preopercular bone which, in spite of its name, does not form part of the opercular series.

The upper mandible joins the suspensorium by means of the palatine. The resulting joint is reinforced by the prevomer and the ethmoid. The lower mandible is attached to the quadrate by the angular, mostly in primitive fishes, or by the articular, in modern ones. In primitive bony fishes, the posterior arm of the suspensorium leans backward, but in modern ones, it attains a vertical position or is inclined forward.

Fig: 26
Bibl: Gregory (1904); Hofer (1945); Schmalhausen (1923)

346. Suture

Fr: suture
Ger: Naht
Lat: sutura (pl: suturæ)
Rus: ШОВ (pl: ШВЫ); КÓСТНЫЙ ШОВ
Sp: sutura

Any joint between two bones consisting of very thin connective tissue, such that there is no movement between them. In Gadidae, the epihyal (= posterohyal) and its ventral part (= anterohyal) are joined on the medial side by a zigzag suture.

Fig: 32

347. Symphysis

Fr: symphyse
Ger: Verwachsung; Symphyse
Lat: symphysis (pl: symphyses)
Rus: сращéние (pl: сращéния); сочленéние (pl: сочленéния);
        симфи́з (pl: симфи́зы)
Sp: sínfisis

Name given to the cartilaginous joint in which two bony surfaces are firmly united by fibro-cartilaginous tissue, as in the mandibular articulation.

348. Symplectic   *

Fr: symplectique
Ger: Symplecticum
Lat: os symplecticum (pl: ossa symplectica)
Rus: СИМПЛÉКТИКУМ
Sp: simpléctico

Paired bone of endochondral origin, that joins the quadrate with the hyomandibular and at the same time supports the interhyal. It is always associated with the hyostylic type of mandibular suspension.

The symplectic is lost in Siluroidei and in anguilloids (Greenwood *et al.* 1966). In *Acipenser*, the symplectic remains cartilaginous during the adult stage.

From its embryonic development the symplectic seems to be a process of the hyomandibular, to which it is sometimes fused. Fused together, these two bones (the symplectic and the hyomandibular) form a compound structure, the *hyosymplectic,* which remains cartilaginous in the adult stages of some fishes.

Figs: 1 D, 3, 13, 25 A, and 30

349. Syncranium

Fr: syncrâne
Ger: Syncranium
Lat: syncranium (pl: syncrania)
Rus: синкра́ниум
Sp: sincráneo

The skull of bony fishes is made up of three main units added one to another through the evolution of fishes, each maintaining its independence. These units are the neurocranium, the branchiocranium and the dermocranium. The anatomical and functional complex made up of these three units is called the *syncranium.*

See   Skull

350. Tabular bones   *

Fr: os tabulaires
Ger: tabular-Knochen; Tabularia; paarige Deckknochen
Lat: ossa tabularia (sing: os tabulare)
Rus: пласти́нчатые (= табли́чные) ко́сти [sing: п-тая
        (= табли́чная) кость]
Sp: huesos tabulares

A group of one to several pairs of dermal bones found in the occipital region, either on the nape or above the opercular membrane, and related to the supratemporal sensory canal (Berg, 1940). In some actinopterygians (*Amia, Lepisosteus,* etc.), they are large, while in other bony fishes they are thin and very small, and consequently often overlooked in dissections.

Syn: scalebones (Gregory, 1933, [1959]); supratemporals (Owen,1866; Ridewood, 1904; and Starks, 1901); extrascapulars; cervicals; nuchals; postparietals.
Figs: 3 and 14 B

351. Tail

Fr: queue
Ger: Schwanz
Lat: cauda (pl: caudæ)
Rus: хвост (pl: хвосты)
Sp: cola

1. That part of the fish body extended between the anus and the tip of the longest caudal rays.
2. In popular parlance, this term applies also to the caudal fin.

Fig: 6

352. Tectospondylous vertebra   *

Fr: vertèbre tectospondyle
Ger: tectospondyler Wirbel
Lat: vertebra tectospondyla (pl: tectospondylæ vertebræ)
Rus: тектоспондильна́льный позвоно́к (pl: т-ные позвонки́)
Sp: vértebra tectospondila

The type of elasmobranch vertebra in which the calcification of cartilage extends both to the arches and to the centrum, resulting in an arrangement of alternating layers. The basking shark (*Cetorhinus*) and the angel shark (*Squatina*) have tectospondylous vertebrae.

Fig: 35 A

353. Teeth   *

Fr: dents
Ger: Zähne (sing: Zahn)
Lat: dentes (sing: dens)
Rus: зу́бы (sing: зуб)
Sp: dientes

Teeth are hard structures attached to the bones of the mandibles, mouth and gills through dermal plates. Teeth appeared for the first time in fish evolution in the mandibles of placoderm fishes. It is recognized that teeth evolved from placoid scales when they invaded the buccal and pharyngeal membranes.

The teeth of fishes present a variety of shapes generally related to their diet. The most common shapes are cardiform, villiform, incisiform, caniniform, and molariform (see these terms). Special modifications are found in the parrotfishes (Scaridae), in which the lower pharyngeal teeth are fused together into a plate of molariform shape, while the upper teeth are free. In this case, the mandibles present small teeth of acrodont or pleurodont type that fuse to them, resembling a parrot's beak. The presence of teeth of similar shape in a fish reflects a uniform diet and is described as *homodonty*, a condition typical of some sharks (*Squalus*). A variety of food in the diet, on the other hand, produces a variety of specialized teeth in the same species, condition known as *heterodonty*.

Teeth are classified and named according to the dermal bones on which they are implanted: premaxillary and maxillary teeth in the upper mandible; the mandibular teeth associated with the dentary; and the vomerian, palatine, ectopterygoid, endopterygoid, and parasphenoid teeth, found in the buccal cavity. In some teleosts (Salmonidae) there is a dermal plate (see Glossohyal) associated with the tongue that bear sharp, prominent teeth. Pharyngeal teeth are found associated with the bones which support the branchiae, by means of an intermediate dermal plate.

Structure of teeth

The teeth of chondrichthyan fishes, joined to the palatoquadrate bars and to Meckel's cartilages by connective fibres, consist of the following four well-defined structures:

1.   the *base* of the tooth, a wide and anatomically independent plate that was later incorporated into the tooth during the evolutionary process. The basal zone cannot be

considered a root. This term is reserved for that part of the tooth embedded in the mandibles of reptiles and mammals.

2. the *body* of the tooth, made up of orthodentine (*Carcharhinus*) or osteodentine (*Lamna*). These materials are deposited by the odontoblasts of the dermis;

3. the *pulp cavity*, filled with special connective tissue cells (*odontoblasts*), blood vessels, and partially occupied by osteodentine; and

4. the *outer layer*, made up of vasodentine, vitrodentine or another enamel-like component. This layer, of ectodermal origin, results from the secretory activity of the *amelocytes* of the epidermis.

Ganoid fishes have teeth, except the sturgeons, that only have them during the larval stages. The teleostean tooth, which lacks a basal plate, is composed of three parts similar to the last three described for those of chondrichthyans. They consist of similar components and their development follows the same plan. The Cyprinidae lack teeth in the mouth but have from one to three rows of them in the fifth ceratobranchial. North American species have only one or two rows of teeth, while in the rest of the world, there are some species with three rows of pharyngeal teeth.

Implantation of teeth

According to the way teeth are anchored to the bone, they can be classified as *acrodont, pleurodont,* or *thecodont.* (See these terms).

Importance of fish teeth in biological and archaeological studies

Teeth can be very useful in identifying fish species and in providing information on their diet and habitat. Among the most valuable teeth for the archaeologist are those of sharks and rays, because of their size and shape. Olsen (1971) states that "some teeth are drilled at their bases an may have been worn as personal ornaments; or they may represent part of a cutting edge, such as saw-toothed weapons made of shark teeth secured to wooden blades".

Many teleosts have teeth that can be identified by shape, but frequently their small size poses a problem in their recovery during excavation. Most prominent and characteristic among fish teeth are the pharyngeal teeth of Cypriniformes (minnows and carps), because their peculiar number, shape, and arrangement on the fifth branchial arch make their identification unmistakable.

Fig: 12
Table 4
Bibl: Applegate (1965); Butler and Josey (1978); Eastman and Underhill (1973); Kerr (1958); Miles (1967); Peyer (1968); Prince (1893); Reif (1978)

354. Teleosts

Fr: Téléostéens
Ger: Teleosteer
Lat: Teleostei
Rus: костйстые рыбы (sing: костйстая рыба)
Sp: Teleósteos

Superorder comprising those actinopterygian fishes whose skeleton ossifies to a greater degree than that of chondrosteans and holosteans.

See   Classification of Fishes
Syn: Teleosteans
Tables 2 and 3

355. Tendon

Fr: tendon
Ger: Sehne (pl: Sehnen)
Lat: tendon (pl: tendines)
Rus: сухожи́лие (pl: сухожи́лия)
Sp: tendón

A tough cord or band of specialized white fibrous connective tissue by which a muscle is attached to the periosteum of a bone, or to the perimysium of a muscle. Tendons serve to transmit the force exerted by a muscle.

356. Terminal vertebra

Fr: vertèbre terminale
Ger: Terminalwirbel; letzer Wirbel
Lat:
Rus: коне́чный (=концево́й) позвоно́к
Sp: vértebra terminal; última vértebra

Gosline (1960) proposed this term to designate the last complete centrum of the vertebral column. Unfortunately, this clear definition has limited value, since the vertebrae so named in different species of fishes are not homologous. The problem arises from the fact that one or more preural vertebrae (PU) are fused to one or more ural vertebrae (U) during juvenile and adult development, and consequently the *ultimate vertebra* can be any one of them. This makes the concept of ultimate vertebra useless for studies of fish evolution and systematics.

Syn: ultimate vertebra
Fig: 13 and 37

357. Thecodont teeth

Fr: dents thécodontes
Ger: thekodonte (= thecodonte) Zähne
Lat:
Rus: текодо́нтные зу́бы (sing: текодо́нтный зуб)
Sp: dientes tecodontos

Name given to teeth implanted in a cavity (*alveolus*) in the bone, similar to mammalian tooth implantation. Thecodont teeth are rare in fishes, since fish teeth lack a true root. The only exception is the rostral teeth of sharks and sawfishes of the families Pristiophoridae and Pristidae, respectively.

See   Rostral teeth

358. Thoracic fishes

Fr: poissons thoraciques
Ger:
Lat: pisces thoracici
Rus: груднопёрые ры́бы
Sp: peces torácicos

A group of bony fishes, proposed by Linnaeus in 1758, to include those fishes whose pelvic fins are attached to the body below the pectoral fins, as in the lumpfish (*Cyclopterus*) and flat fishes.

Syn: subbrachial fishes (Cuvier)
Table 3

359. Total length

Fr: longueur totale
Ger: Totallänge
Lat: longitudo maxima
Rus: общая (= абсолютная) длина̄
Sp: longitud total

One of the main interests in archaeological faunal studies is to estimate the live size of the fish from the remains found (bones, teeth, scales or otoliths). At the same time, more and more zooarchaeologists are realizing the value of obtaining as much biological information as possible from fish remains (size, age, sex, weight, dietary value, season of capture) in order to gain an understanding of the cultural activities (fishing, diet, customs, migrations) and the ecological conditions (seasonality, climate) of the fish and of past human communities.

Because fish length is the easiest feature to measure, it has been most frequently related to the age and the weight of the fish in biological research, than any other measurement.

Since there are no written records of the biological characteristics of the fish remains, one must relate subfossil bone dimensions to present day fish size. Size and growth data of modern fish are available from numerous biological publications. The problem with this sort of data is that it has been obtained using different methods, and therefore it is difficult to compare the results.

Fish growth depends on the availability of food, which depends on the environmental conditions of the area. Food availability is responsible for the variability in fish size at any given age. For example, six-year old Labrador codfish (area 2J of ICNAF) is some 10 cm shorter (Rojo, 1959) than those from Greenland (area 1D). For this reason, conversion factors and equations used to calculate fish live size should be taken from an area as close as possible to the archaeological site and/or from a place as ecologically similar as possible.

The concept of fish length, obvious in principle, can lead to gross errors in the estimation of fish size, if fish length has not been defined accurately. Several "fish lengths" are, or have been, used in biological research. Therefore archaeologists should check first which length is the most appropriate for their specific purposes.

Fish length has been taken in many different ways according to the species characteristics and the particular research practice. The most common lengths taken are the following: total length, standard length, and fork length (see these terms).

Total length is the distance taken in a horizontal, straight line from the tip of the snout to the end of the longest caudal fin ray. The total length can be taken in several ways, depending on the species and the purpose of the work. Some of the most common ways of taking the total length are:

a)   leaving the caudal lobes spread in their natural position, and taken the distance to the line joining their tips or, if unequal in length, to the tip of the longest lobe (AG in figure 7). In either case the total length is also called *total normal length* or *natural tip length*;

b)   squeezing the tail lobes or only the longest one towards the middle line (A'I in figure 7) This length has been called *maximum length* or *extreme tip length;* and

c)   drawing an imaginary line joining the tips of the caudal lobes and measuring up to its middle point. The length so obtained is called *total auxiliary length* or *bilobular length* (A'H in fig. 7).

In all cases, the measurement is taken between the projections on a horizontal plane of the points described, and should not be taken over the curve of the body.

From a mathematical and even from a biological point of view, it seems that there is no clear advantage in using any one of the measurements defined. A similar conclusion has been drawn when the total length is calculated using standard or fork lengths. Papers have been published defending or rejecting the value of any one of these measurements. Royce (1942) found, however, that total length measurements are more representative of the fish weight than standard length for perch (*Perca flavescens*) and lake trout (*Cristivomer namaycush*). The most important criteria applicable in each case is to maintain the consistency in the method used, to follow the practice of the specialists for each fish group, and to describe exactly the procedure used.

The discrepancy in the values of the length used by different authors can be resolved by calculating the regression equations between each pair of measurements in question (total length vs. standard length; total length vs. fork length; fork length vs. standard length), for each species, location, and season. In this way it is possible to compare published data taken with different techniques.

Fig: 7
See   Fork length, Standard length, and Head length

360. Tropibasic skull

Fr: crâne tropybasique
Ger: tropibasischer Schädel
Lat:
Rus: тропибазáльный   чéреп
Sp: cráneo tropibásico

Skull characterized by the narrow distance between the trabeculae, which fuse early in development by means of a transverse commissure, the *trabecula communis*. The common trabecule forms later the nasal and the interorbital septa. As a consequence, the eyes are set closer to each other than in more primitive fishes. This condition is found in teleosts and represents a more advanced stage in the evolution of fishes. The primitive condition is that of the platybasic skull.

Syn: tropitrabic skull
See   Platybasic skull

361. Trunk

Fr: tronc
Ger: Rumpf
Lat: truncus (pl: trunci)
Rus: ствол; тýловище
Sp: tronco

That part of the fish body between the posterior border of the opercular membrane and the vent.

Fig: 6

362. Ural vertebrae *

Fr: vertèbres urales
Ger: urale Wirbel
Lat: vertebræ urales (sing: vertebra uralis)
Rus: уральные позвонки (sing: уральный позвонок)
Sp: vèrtebras urales

Nybelin (1963) defined ural vertebrae as those to which hypural bones are attached.

See Hypurals
Fig: 37

363. Urodermals *

Fr: urodermaux (sing: urodermal)
Ger: Urodermalia
Lat: ossa urodermalia (sing: os urodermale)
Rus:
Sp: urodermales

The numerous and paired thin bones of dermal origin, located back of the caudal skeleton of primitive teleosts. They were called previously *uroneurals* by Regan (1910). Nybelin (1963) interpreted them to be modified scales sunk in the musculature and renamed them *urodermals*.

See Caudal skeleton

364. Urohyal *

Fr: urohyal
Ger: Urohyale
Lat: os urohyale (pl: ossa urohyalia)
Rus: заднеподъязычная кость
Sp: urohial

A median endochondral bone resulting from the ossification of the cartilage located between the two sternohyoid muscles in teleosts (Ridewood, 1904). In spite of its name, it does not belong to the hyoid arch. This bone is missing in primitive fishes, such as *Lepisosteus* (Jollie, 1984).

Syn: parahyoid; clidost (Chabanaud); episternal (Geoffroy St-Hilaire); interclavicle (Cole and Johnston).

Figs: 25 A and 34 B
Bibl: Kusaka (1974)

365. Uroneurals *

Fr: uroneuraux (sing: uroneural)
Ger: Uroneuralia
Lat: ossa uroneuralia (sing: os uroneurale)
Rus: уроневралии

Sp: uroneurales

Paired endochondral bones located anteriorly above the urostyle and present in advanced holosteans and in many teleosteans. The largest and most advanced pair in Salmonidae is also known as *caudal bony plate*.

See   Caudal bony plate and Caudal skeleton
Fig: 37

366. Urostyle   *

Fr: urostyle
Ger: Urostyl
Lat: urostylus (pl: urostyli)
Rus: уростйль
Sp: urostilo

1. The last segment of the vertebral column. It is a slender, pointed rod of cartilage or bone representing the fusion of several vertebrae (Whitehouse, 1910).

2. The urostyle is "the fanlike tail segment articulated with the last true vertebra" (Clothier, 1950), or in other words, all the vertebral elements which fused into one piece, following the last undoubted centrum. The urostyle, in its most complex development, is made of preural centra, ural centra, epurals and hypurals. The urostyle is considered *one unit* in the total vertebral count (Hollister, 1936). Many European biology studies do not include the urostyle in the vertebral count, while American works normally do. This difference has to be taken into account when comparing data from both sources.

Syn: opisthure
Fig: 13

367. Utricle

Fr: utricule
Ger: Utriculus; Utrikulus
Lat: utriculus (pl: utriculi)
Rus:
Sp: utrículo

The utricle, the uppermost of the three chambers of the membraneous labyrinth, is closely associated with the semicircular canals. In cartilaginous fishes, it is filled with endolymph interspersed with minuscules particles of calcium carbonate, called *otoconia*. In Teleosts, the otoconia are replaced by the *utriculith* or *lapillus* , which is generally the smallest of the three otoliths. Because of its small size, it is required water-sieving methods to retrieve them in the digs.

See   Membraneous labyrinth
Fig: 42

368. Vertebrae   *

Fr: vertèbres
Ger: Wirbel

Lat: vertebra (pl: vertebræ)
Rus: ПОЗВОНКЍ (sing: ПОЗВОНŌК)
Sp: vértebras

Each of the metameric units into which the vertebral column is divided. Like many other organs of the body, the vertebrae have had a long and complex evolutionary history, repeated in their embryonic development. As a consequence, the concept of vertebra, clear and distinct when referring to that of teleosts and tetrapods, is more imprecise when applied to the incomplete vertebrae of cyclostomes and primitive fishes.

The absence or presence of vertebrae has been used to divide the animal kingdom into two large divisions: the Invertebrates and the Vertebrates. This classification is somewhat arbitrary since there are some animals (*Amphioxus*) without vertebrae, and others with incomplete vertebrae (lamprey) that resemble vertebrates in many other features.

The embryogeny of the fish shows the many elements that intervene in the formation of a vertebra. These are:

a)   the neural arch formed by the *interdorsal* and *basidorsal* of each side, although to a different degree in the various fish groups;

b)   the *neural spine*, fused to the neural arch;

c)   the vertebral body or *centrum*, formed by the union of two *pleurocentra* and two *intercentra* , or parts thereof;

d)   the *hemal arch*, formed from the more-or-less complete fusion of the *interventral* and the *basiventral;* and

e)   the *hemal spine*, which is added to the hemal arch.

The phylogenetic and embryonic order of the three most important structural units forming the vertebrae is, first, the neural arch; second, the hemal arch, and third, the centrum or body of the vertebra.

Owing to the complex character of the vertebrae, each one of the larger taxonomic group of PISCES shows a wide range of structures in the development of the vertebral column.

Comparative anatomy of the vertebrae

Lamprey vertebrae have no centra, since the notochord persists as a continuous unit. However, in the caudal region, there are some cartilages arranged in an inverted 'V' pattern above, and in a 'V' position below, representing the first neural and hemal arches, respectively.

The elasmobranch vertebra consists of three elements: the body, the neural arch and the hemal arch. The body or centrum has a cylindrical core of notochordal tissue different in diameter in each group. This core is surrounded by a ring of cartilage, known as *chordacentrum*, to which the bases of the neural and hemal arches abut. The cartilaginous arch and its base are known collectively as the *arcocentrum*. In the space between the bases that surround the chordacentrum, there is another interrupted ring of cartilage, called *autocentrum*. Each one of these units varies in size and structure in the different species (Fig. 56 A).

A calcification process soon takes place, resulting in the hardening of elasmobranch vertebrae. According to White (1937) each one of the three cartilaginous units is reinforced by the deposition of calcium, which takes the form of rings, radii, or a pattern known as the "Maltese Cross". According to the units involved -- chordacentrum, autocentrum or arcocentrum -- and the calcification pattern, the vertebrae can be classified as *cyclospondylous* (when only the chordacentrum is calcified), *asterospondylous* (when the calcification extends to the chordacentrum, in the shape of a simple ring with or without radial plates, and also to the autocentrum), and *tectospondylous* (when the calcification invades all three sections). The neural arches join each other, forming a continuous canal, that protects the spinal cord. Likewise, the hemal arches offer protection to the caudal artery and vein.

In holocephalans (*Chimaera*), the vertebrae are aspondylous, i. e. without a centrum, but with neural and hemal arches well developed. The notochord remains continuous and uniform in diameter along its whole length.

In chondrosteans (sturgeons and paddlefishes), the notochord persists as a cylindrical, uninterrupted structure without constrictions.

In holosteans (*Amia*), there is a complete record of the sequence of the formation of a vertebra. The various units -- pleurocentrum, intercentrum, basidorsal, basiventral, interdorsal, and interventral -- ossify and remain independent during the embryonic and juvenile stages (*dyssospondyly* or *dyssospondylous condition*). Some of these elements fuse later, producing vertebrae with two centra or bodies (*diplospondyly* or *diplospondylous condition*) located in the caudal region. But in the anterior section of the vertebral column, both centra join together and form the typical vertebra with all their elements fused (*holospondyly* or *holospondylous condition*).

In teleosts, the juvenile and adult vertebrae are also monospondylous with anterior and posterior concave facets, giving when viewed in a lengthwise section, the appearance of an hourglass. Because of the presence of this double cavity, they are called *amphicoelous vertebrae*. The spaces between two vertebrae are filled with notochordal tissue that continues from one vertebra to the next through a canal pierced in the centrum of the vertebrae.

Above the neural arches, a new ossification of the dorsal septum (a structure which separates the two lateral masses of musculature) joins the two branches of the neural arch, forming the *neuracanth* or *neural spine*. In a similar way, the hemal arch extends into the *hemacanth* or *hemal spine*.

Vertebrae are the most numerous fish bones found in archaeological sites. They are also the easiest to be recognized as such, although in some cases they are difficult to be assigned to the right species. The simplest method to identify fish vertebrae is to compare them with a collection of fish bones, but unfortunately such a collection is not always available. Radiographic techniques have been applied successfully for the recognition of fish vertebrae by Cannon (1988); Desse and du Bruit (1970-1971); Desse and Desse (1976, 1983), and Rojo (1989).

Vertebrae from sharks, rays, and fishes of the families Scombridae, Lophiidae, and Gadidae have also been used to reconstruct the live size of the fish (Rojo, 1987) and to estimate their age.

　　　See　　Vertebral column
　　　Figs:　13, 35, 36, and 37
　　　Bibl:　Casteel (1976); Fahy (1972); Ford (1937); Ridewood (1921); Rojo (1987); Schaeffer (1967); Shute (1972);

369. Vertebral column

　　　Fr:　　colonne vertébrale
　　　Ger:　Wirbelsäule
　　　Lat:　columna vertebralis (pl: columnæ vertebrales); spina dorsalis (pl: spinæ dorsales)
　　　Rus:　позвоночный столб
　　　Sp:　　columna vertebral

　　　The vertebral column of bony fishes extends from the skull to the caudal fin and consists of more or less complete cartilaginous or bony units, the vertebrae. The vertebral column supports and protects the spinal cord by means of the centra and the neural arches, a function which it carries out more efficiently than the notochord which it replaces. Moreover, the vertebral column serves as an attachment for the powerful muscles that form the *dorsalis trunci,* which are the organs primarily responsible for fish locomotion.

The vertebral column can be divided into several regions, although the boundaries between them in fishes are not as well marked as in the remaining vertebrates. Two main regions can be established at first glance, separated by the position of the vent: the anterior, *abdominal* or *precaudal* (sometimes also called *thoracic*) and the posterior, the *caudal*.

The abdominal vertebrae lack hemal arches, while in the caudal vertebrae, the parapophyses join at their ends, forming an arch. The abdominal region can, in some families (Gadidae), be in turn, divided into two sections: the *cervical* or *prethoracic region,* whose vertebrae lack parapophyses, and the abdominal, with well-developed parapophyses.

The total number of vertebrae can be expressed using three numbers, which represent the cervical (= prethoracic), abdominal (= thoracic), and caudal vertebrae, respectively. In this last group, many North American studies include the urostyle as a caudal vertebra. European studies usually omit this last bone. Therefore, when reporting the number of vertebrae, it should be mentioned whether the urostyle has been included or not. The arithmetic expression of these values is known as the *vertebral formula ,* which for a sample of 82 Atlantic cod off the coast of Nova Scotia (Canada) was found to be

Prethoracic	Abdominal	Caudal	Total
C_4	$T_{(13-16)}$	$C_{(33-37)}$	$T_{(51-55)}$

In the calculation of the MNI in archaeological work, it is important to know the variability in the number of vertebrae in a particular species from a particular area, since the number of vertebrae is a meristic character determined both by inheritance and environment. For this reason, it is more accurate to calculate the MNI using the mean number of vertebrae for each fish and place, rather than the actual minimum or maximum number of the sample.

Fig: 13
Bibl: Ford (1937); Gómez-Larrañeta (1958); Hubbs (1922); Mookerjee (1936); Springer and Garrick (1964)

370. View

Fr: vue
Ger: Ansicht
Lat: norma (pl: normæ)
Rus: ВИД
Sp: vista

The aspect or appearance that a bone presents when viewed from a particular position.

371. Villiform teeth

Fr: dents en soie
Ger: zottenförmige Zähne
Lat:

Rus: ворсинковидные (=щетиновидные) зубы [sing:
        ворсинковидный (=щетиновидный) зуб]
Sp: dientes viliformes

    Villiform teeth are thin and long, either packed densely, as in the needlefish (*Belone*), or isolated, as in the lampreys *Tetrapleurodon* and *Eudontomyzon*.

372. Visceral skeleton

Fr: squelette viscéral
Ger: Visceralskelett
Lat:
Rus: висцеральный скелёт
Sp: esqueleto visceral

    The visceral skeleton, considered in its widest sense, comprises all the cartilaginous and bony elements related either to the respiratory apparatus (gills) or to the anterior part of the digestive system (mouth and pharynx). Both systems, respiratory and digestive, are closely interrelated in fishes. Some of their bony elements have a double function, that of supporting both the branchial filaments (respiratory activity) as well as the gill rakers, teeth and dental plates (feeding function).
    In bony fishes, the visceral skeleton consists of three units: *the mandibular arch,* which forms the primary mandibles; the middle unit, *the hyoid arch*; and the posterior one, formed by the U-shaped *branchial arches*.
    Since all these anatomical units are located in the fish head (see Head), they conjointly receive the name of *splanchnocranium* or *viscerocranium*. The lower section of the hyoid arch, together with the branchial arches, receives the not-so-well-known name of *hyobranchium;* while the branchial skeleton proper is known as *branchiocranium*.

373. Vomer

Fr: vomer
Ger: Vomer
Lat: os vomere
Rus: сошник
Sp: vómer

    Paired dermal bone found in most primitive fishes. In the remaining actinopterygians both bones fuse into one vomer, forming the roof of the palate in its anterior part. It is covered dorsally by the ethmoid and articulates firmly with the parasphenoid. Frequently the vomer bears teeth, either of a caniniform (*Amia*) or molariform type (*Anarhichas*).

See   Prevomer
Figs: 14 B and 19

374. Weberian apparatus   *

Fr: appareil de Weber
Ger: Weberscher Apparat
Lat: apparatus weberianus (pl: apparati weberiani)
Rus: Вёберов (=вёберовский) аппарат
Sp: aparato de Weber

Weber (1820) was the first to describe a set of bones in fishes which he considered to be related to the sense of hearing and to be homologous to the mammalian ear ossicles. Consequently, he assigned them the names *malleus, incus,* and *stapes.* The anteriormost, he called *claustrum* because is the most deeply set in the body.

Bridge and Haddon (1889) showed that these bones originate from vertebral elements and from the first ribs, and therefore they could not be homologous to the ear ossicles. In order to correct the old interpretation, they proposed the new term *Weberian apparatus* for the complex unit which can be subdivided into the four following parts:

a) the *auditory unit (pars auditum)* consisting of a paired chain of four ossicles, named in anteroposterior sequence, *claustrum, scaphium, intercalarium,* and *tripus,* and known collectively as Weberian ossicles or *ossa auditoria* ;

b) the *supporting unit (pars sustentaculum)* formed in most Ostariophysans by two transverse plates extending downward from the fourth vertebra, leaving between them a circular space for the passage of the aorta;

c) the fused *cervical centra* (the second and third in many species) with their transverse processes; and

d) the *neural complex* , resulting from the modification of the neural arches and spines.

The orders Cypriniformes and Siluriformes were previously grouped as Ostariophysi, because of the relationship of these bones with the gas bladder. The differences in size and shape of these bones have a species-specific value in taxonomy within the ostariophysan group. The original names, proposed by Weber, were replaced by Bridge and Haddon with new ones, as follows:

1. the malleus of Weber is now called *tripus.* This last name derives from the presence on the bone of three processes: one anterior, joined to the intercalarium bone by a ligament; another median, that articulates with the third vertebral centrum; and a third, posterior, ending in the transformator process which rests on the anterior wall of the gas bladder;

2. the incus of Weber is the *intercalarium;*

3. the stapes of Weber is now called *scaphium* because of its spoon or skiff shape; and

4. the anteriormost, the *claustrum* of Weber, retained its name, since there is no confusion possible with any of the mammalian ear ossicles.

Müller (1853) and Thilo (1908) proposed new names which have not been generally accepted. Many anatomists have studied the development of the Weberian apparatus, but unfortunately their opinions show an astonishing discrepancy in the interpretation of their embryological origins.

After the decomposition of the fish on the ground, the Weberian apparatus breaks into pieces, but even in this condition it is possible an accurate identification of the same, in archaeological work.

Fig: 36
Bibl: Krumholz (1943); Rosen and Greenwood (1970)

375. Weight

Fr: poids
Ger: Gewicht
Lat: pondus
Rus: вес
Sp: peso

The weight is a measurement of the mass or force with which a body responds to gravity, generally expressed, in fish studies, in metric units (grams or kilograms) depending on the size of the specimen or organ weighed.

In fishery biology, when studying the productivity of a body of water, the fish weight is taken when the fish is still alive or immediately after death. Knowing the weight, it is possible to calculate the biomass produced per unit of time, volume, or surface. But in works with a commercial aim (market and dietary studies, etc.) it is useful to calculate the weight of that part of the fish body which is utilized for industrial processing or human or animal consumption.

Two weights are used in commercially oriented studies: the total and the dressed weights. The former, taken immediately after the death of the fish, is used more often because it is easily obtained. It has however, the disadvantage of being very variable in relation to the length of the fish. Among the factors that affect it are: the weight of the gonads during the maturation and spawning seasons; the weight of the liver during the summer feeding season; the weight of the stomach which varies with the time of day; the sex, especially in gravid females; the consistency of the flesh, depending on age; and the state of health of the fish. All these factors should be taken into consideration in faunal studies, when estimating the live weight of the fish, since the total food value obtained varies widely when only one or a few factors are used in its estimation.

Another drawback in estimating weight, especially total weight, is that there are very few studies directly relating the bone size to fish weight. Most authors use an indirect approach, estimating first the length of the fish from the bone size and then, the fish weight using the fish length calculated previously. This two-step procedure results in a larger margin of error than when relating directly the bone size to the fish weight. The correlation coefficient for the relationship between bone size to fish weight is always much smaller than that obtained for bone size to fish length. Biological works deal only with the relationship fish length to fish weight.

Dressed weight, taken after evisceration of the specimen, is more constant in its relationship to bone size and fish length, than is total weight. It also better reflects the dietary value of the fish consumed. For small fish, naturally, the difference between total and dressed weight is usually not significant.

When the fish is processed before being sent to market, one can calculate the weight loss. Codfish, for example, is sold in different conditions, such as fresh gutted, gutted "head on", "green" dry, and salted. Then, it is necessary to know the weight loss after each one of the different stages of preparation: removal of the head, evisceration, salting, and drying. A simple calculation gives a factor for each one of these processing stages.

The conversion factor is a number that, multiplied by the weight after each process, gives the original weight. For the cod of the Northwest Atlantic, a conversion factor of 3 was accepted as representative of the weight reduction between the weight *in vivo* and the salted product (FAO, 1970). In this case, the loss is approximately 2/3 of the original weight. The conversion factors between the live fish and the marketable product should be calculated for each commercial species in each fishery area for each season. The values obtained can have a wide variation, due to the biological characteristics of the fish (growth rate, age at maturation, gonadal development) and the environmental factors corresponding to each region in a particular year.

This somewhat inaccurate method has been applied in archaeological faunal studies to fish and mammal remains, with dubious results.

The archaeologist interested in fish faunal studies should therefore prepare direct mathematical relationships between bone size and fish weight, either total or dressed, when the size of the sample or the nature of the study warrants them.

This relationship is best expressed by an exponential equation of the type

$$W = a \times L^n$$

which is usually expressed logarithmically, to simplify its application. The new equation is

$$\log. W = \log. a + b (\log. L)$$

where $\underline{W}$ is the fish weight and $\underline{L}$, its total, standard, or fork length. The constant $\underline{a}$ varies for each species, sex, population, and time in which the fish lived. The exponent $\underline{b}$ has a value between 2 and 4, although it oscillates in most cases around 3.

  An exact value of 3, rarely found, indicates an isometric growth for the fish length and weight. Values above and below this figure represent allometric growth. When $b > 3$, it means that the fish grows faster in weight than in length, as is the case of fishes with globular and truncate bodies. When $b < 3$, the body elongates rapidly, while the weight increases more slowly.

  In archaeological work, before calculating the weight or calories of the fish found, one must establish the proportion of body fish mass taken as food. Many fish organs discarded in our highly industrialized societies (ovaries, liver, head meat, small bones) are utilized by economically poor communities and were also most likely eaten by primitive peoples. Even whole fish, including the skin and cartilage of small sharks, rays, and chimaeras, are now sometimes dried and processed, to be consumed as consommé.

Bibl: Casteel (1974a); Cirile and Quiring (1940); LeCren (1951)

7. BIBLIOGRAPHY

ADAMS, L. A.
>   1940   Some characteristic otoliths of American Ostariophysi. *J. Morph.* 66 :
>   497- 527

AFFLECK, K. J.
>   1950   Some points in the function, development and evolution of the tail in
>   fishes. *Proc. Zool. Soc.* London. 120 : 349 - 368

AGASSIZ, L.
>   1833   Recherches sur les poissons fossiles. Neuchatel. 1: 61-90.

ALEEV. Y. G.
>   1963   *Function and gross morphology of fish. Israel Prog. for Sci. Trans.*
>   1969 128 pp.

ALLIS, E. P. Jr.
>   1897   The cranial muscles and cranial and first spinal nerves in *Amia calva.*
>   *J. Morphol.* 12 (3): 487- 808.

ALLIS, E. P. Jr.
>   1903   The skull and cranial and first spinal muscles and nerves of *Scomber
>   scomber. J. Morphol.* 18 : 45-328. 68 figs.

ALLIS, E. P. Jr.
>   1922   The cranial anatomy of *Polypterus,* with special reference to *P.
>   bichir. J. Anat.* 56 (3-4): 189-294

APPLEGATE, Sh. P.
>   1965   Tooth terminology and variation in sharks with special reference to
>   sand shark, *Carcharias taurus.* Rafinesque. *L. A. Co. Mus. Nat.
>   Hist. Contrib. to Sci.* 86 : 1 -18

ARITA, G. S.
>   1971   A re-examination of the functional morphology of the softrays in
>   teleosts. *Copeia.* 1971 (4) : 691-697

ATZ, J. W.
>   1968   Dean bibliography of fishes. *Amer. Mus. Nat. Hist.* 512 pp. New
>   York.

ATZ, J. W.
>   1969   Dean bibliography of fishes. *Amer. Mus. Nat. Hist.* 853 pp. New
>   York

AUMONIER, F. J.
>   1942   Development of the dermal bones in the skull of *Lepidosteus osseus.
>   Q. J. Microsc. Sci.* 329 : 1-8.

BAILEY, R. M.; J. E. FITCH; E. L. HERALD; E. A. LACHNER; C. C.
    LINDSEY; C. R. ROBINS and W.B. SCOTT.
>   1970   A List of Common and Scientific Names of Fishes of the United
>   States and Canada. *Amer. Fish. Soc. Spec. Publ.* 6 : 1 - 149. Third
>   edition. Washington.

BALON, Eugene K. and D. L. G.
>   1980   *Principles of Ichthyology.* Department of Zool. College of Biol. Sci.
>   Univ. Guelph, Ontario. Canada. 285 pp

BARRINGTON, E. J.
>   1935   Structure of the caudal fin of cod. *Nature.* 135 (85): 270

BEER, G. R. de,
>   1937   The development of the vertebrate skull. *Oxford : Clarendon Press.*
>   552 pp.

BENECKE, Norbert
>   1986   Some remarks on sturgeon fishing in the southern Baltic region in
>   Medieval times. 9-17 pp. In: *Fish and Archaeology. Studies in*

osteometry, taphonomy, seasonality and fishing methods. (Editors) D. C. Brinkhuisen and A. T. Clason. Bar International Series 294

BERG, L. S.
1912   Poissons (Marsipobranchii et Pisces). *Faune de la Russie.* 3 (1): 1-336. (in Russian)

BERG, L. S.
1940   Classification of fishes, both recent and fossil. Moscow. *Akademiia Nauk S.S.S.R.* 5 (2): Part I (in Russian) 87-345; Part II. (in English): 346-511

BERG, L. S.
1958   *System der rezenten und fossilen Fischartigen und Fische.* Berlin. 293 pp.

BERRY, F.H.
1964   Aspects of the development of the upper jaw bones in fishes. *Copeia.* 1964 (2): 375 - 384

BERTIN, L.
1944   Modifications proposées dans la nomenclature des écailles et des nageoires.*Bull. Soc. Zool. Fr.* 69 :198-202.

BERTIN, L. & C. ARAMBOURG.
1958   Systématique des poissons. In: *Traité de Zoologie.* Edited by P. P. Grassé, 13 (3): 1967-1983.

BLANC, M.;P. BANARESCU; J-L. GAUDET and J-C. HUREAU
1971   *European inland Water Fish. A multilingual catalogue. Peces de las aguas continentales de Europa. Catálogo bilingüe.* Fishing News (Books) Ltd. London

BÖKER, H.
1913   Der Schädel von *Salmo salar.* Ein Beitrag zur Entwicklung des Teleostierschädel *Anat. Hefte* 49: 359-396

BONAPARTE, C. L.
1832   *Iconografia della fauna italica per le quattro classi degli animali vertebrati.* Pesci Vol. III. Roma. 41 plates

BOND, C. E.
1979   *The Biology of Fishes.* Philadelphia: W.B. Saunders. 514 pp

BOULANGER, G. A.
1904   A synopsis of the suborders and families of the teleostean fishes. *Ann. Mag. Nat. Hist.* 7 :161-190

BRIDGE, T. W. The cranial osteology of *Amia calva. J. Anat. and Physiol.* 11 : 605-622. 1877 London.

BRIDGE, T. W. & A. C. HADDON.
1889   Contributions to the anatomy of fishes. I. The airbladder and Weberian ossicles in the *Siluridae. Proc. R. Soc. London.* 46:309-328

BRIGGS, J. C.
1966   Zoogeography and evolution. *Evolution.* 20 : 282 - 289

BRINKHUISEN, Dick, C.
1986   Features observed on the skeletons of some recent European Acipenseridae: their importance for the study of excavated remains of sturgeons. 18 33 pp. In: *Fish and Archaeology. Studies in osteometry, taphonomy, seasonality and fishing methods.* (Editors) D. C. Brinkhuisen and A. T. Clason. BAR International Series 294

BRUCH, C. W. L.
1861   Vergleichende Osteologie des Rheinlachses (*Salmo salar*) mit besonderer Berücksichtigung der Myologie nebst einleitenden Bemerkungen über die skelettbildenden Gewebe der Wirbeltiere. Mainz.

BRÜHL, C. B.
1877   *Zur Osteologie der Knochenfische.* 76 pp. 150 figs; 11 tables. Berlin.

BRYAN, A. L.
  1963   An archaeological survey of Northern Puget Sound. *Idaho State Univ. Mus. Occas. Papers.* 11.

BUHAN, P.J.
  1972   The comparative osteology of the caudal skeleton of some North American minnows (Cyprinidae). *Amer. Middl. Nat.* 88 (2) : 484 -490

BULKLEY, Ross, V.
  1960   Use of the branchiostegal rays to determine age of lake trout, *Salvelinus namaycush* (Walbaum). *Trans. Amer. Fish. Soc.* 89 (4) : 344 -350

BUTLER, P. M. and K. A. JOSEY
  1978   *Development, function and evolution of teeth.* New York : Academic Press

CABLE, L. E.
  1956   Validity of age determination from scales, and growth of marked Lake Michigan lake trout. *U. S. Fish Wildl. Serv. Fish. Bull.* no. 107, vol. 57: 1-59

CAILLET, Gregor, M.; M. S. LOVE and A. W. EBELING
  1986   *FISHES: a field and laboratory manual on their structure, identification and natural history.* Wards publishing Company, Belmont, California. 193 pp.

CANNON, Aubrey
  1988   Radiographic age determination of pacific Salmon: species and seasonal differences. *Journal of Fish Archaeology.* 15: 103-108

CANNON, Debbi Yee
  1987   Marine fish osteology: a manual for archaeologists. *Publication (Simon Fraser University. Dept. of Archaeology).*18: 1-133.

CARLANDER, Kenneth, D.
  1950   Growth rate studies of saugers, *Stizostedion canadense canadense* (Smith), and yellow perch, *Perca flavescens* (Mitchill), from Lake of the Woods, Minnnesota. *Trans. Am. Fish. Soc.* 79: 30-42

CARLANDER, Kenneth, D.
  1969   *Handbook of Freshwater Fishery Biology.* 3rd edition. Vol. 1: 752 pp. Life history data on freshwater fishes of the United States and Canada, exclusive of Perciformes. Iowa State Uni. Press. Ames. Iowa.

CARLANDER, Kenneth, D.
  1977   *Handbook of Freshwater Fishery Biology.* Vol. 2: 431 pp. Life history data on Centrarchid fishes of the United States and Canada. Iowa State Univ. Press. Ames, Iowa.

CASIER, E.
  1954   Contribution à l'étude des poissons fossiles de la Belgique. XI. Note additionelle relative à *"Stereolepis"* (*Osorioichthys* nov. nom) et à l'origine de l'interoperculaire. *Meded. K. Belg. Inst. Natuurwet.* 30 (2): 1-12.

CASSELMAN, John, M.
  1974   Analysis of hard tissue of pike *Esox lucius* L. with special reference to age and growth. In: *Ageing of fish.* 13-27(Editor) T. B. Bagenal. Unwin Brothers Limited. Surrey, England.

CASSELMAN, John, M.
  1983   Age and Growth Assessment of Fish from Their Calcified Structures-Techniques and Tools. 1-17 pp. In: *Proceedings on the International Workshop on Age Determination of Oceanic Pelagic Fishes: Tunas, Billfishes, and Sharks.* Eric. D. Prince (Convener and editor) and L. M. Pulos (editor) NOAA Technical Report NMFS 9.

CASTEEL, Richard, W.
  1972   Some archaeological uses of fish remains. *Amer.Antiq.* 37 (3): 404-419

CASTEEL, Richard, W.
  1973   The scales of the native freshwaterfish families of Washington.
          Northwest Sci. 47 (4) : 230 - 238
CASTEEL, Richard, W.
  1974 a A method for estimation of live weight of fish from the size of skeletal
          elements. *Amer. Ant.* 39 (1) : 94 - 98
CASTEEL, Richard, W.
  1974 b Identification of the species of Pacific salmon (genus *Oncorhynchus*)
          native to North America based upon otoliths. *Copeia.* 1974 (2) :
          305 - 311
CASTEEL,Richard, W.
  1974 c On the number and sizes of animals in archaeological faunal
          assemblages. *Archaeometry.* 16 (2) : 238 - 243
CASTEEL, Richard, W.
  1975   On the remains of fish scales from archaeological sites. *American
          Antiquity.* 39 (4): 557 -581
CASTEEL, Richard, W.
  1976   *Fish remains in Archaeology and Palaeo-environmental Studies.*
          180 pp. Academic Press. New York.
CHABANAUD, P.
  1936   Le neurocrâne des Téléostéens dissymétriques après la metamorphose.
          Ann. Inst. Oceanographie. 16 : 223-297
CHAPLIN, Raymond, E.
  1971   *The Study of Animal Bones from Archaeological Sites.*   170 pp.
          Seminar Press. London and New York.
CHAPMAN, W. Mc.
  1941   The osteology and relationships of the isospondylous fish *Plecoglossus
          altivelis,* Temminck and Schlegel. *J. Morphol.* 68 (3): 425-455
CIRILE, G. and D. P. QUIRING
  1940   A record of the body weight and certain organs and gland weights of
          3690 animals. *Ohio J. Sci.* 40 : 219 - 259
CLEMENS, W. A. and G. V. WILBY
  1949   Fishes of the Pacific Coast of Canada. *Fish. Res. Board. Canada. Bull.*
          68 : 1-368
CLOTHIER, C. R.
  1950   A key to some Californian fishes based on vertebral characters. *Calif.
          Depart. Fish and Game.* Fish. Bull 79
COCKERELL, T. D.
  1913   Observations on fish scales. *Bull. U. S. Bur. Fish.* 1912, 32 :
          119 -174
COHEN, D. M.
  1970   How many recent fishes are there? *Proc. Calif. Acad. Sci.* Series 4.
          38 (17): 341-346.
COMPAGNO, L. J. V.
  1973   Interrelationships of living elasmobranchs. In : *Interrelationships of
          Fishes .* New York: Academic Press. 15 - 61
COMPAGNO, L. J. V.
  1977   Phyletic relationships of living sharks and rays. *Amer. Zoologist.* 17 :
          303 - 322
COPE, E. D.
  1890   The homologies of the fins of fishes. *Am. Nat. ,* 24: 401-423
COURTEMANCHE, Michelle and Legendre, VIANNEY
  1985   Os de poissons: Nomenclature codifiée. Noms français et anglais.
          Rapport technique. 06-38. 61 pp.

DAHL, Knut
  1909   The assessment of age and growth in fish. *Intern. Revue der gesamten Hydrobiologie und Hydrographie.* 2 (4-5) : 758-769. Leipzig
DANIEL. J. F.
  1934   *The Elasmobranch Fishes.* Berkeley : Univ. California Press. 3rd Edition
DARLINGTON, P. J. Jr.
  1957   *Zoogeography.* New York: John Wiley and Sons [Freshwater fishes] 127 pp.
DEAN, B.
  1916   A bibliography of fishes. Vol. I A-K. 718 pp. *Amer. Mus. Nat. Hist.* New York.
DEAN, B.
  1917   A bibliography of fishes. Vol. II. L-Z. 702 pp. *Amer. Mus. Nat. Hist.* New York
DEAN, B.
  1923   A bibliography of fishes. Vol. III. 709 pp. *Amer. Mus. Nat. Hist.* New York.
DEGENS, E. T.; W.G. DEUSER and R.L. HAEDRICH
  1969   Molecular structure and composition of fish otoliths. Collected reprints. Woods Hole Ocean Agr. Inst. Contrib. 2214 (reprinted *Marine Biology* 2 (2) : 105-113
De KAY, J. E.,
  1842   The American hake. In : *New York Fauna.* Natural History of New York. New York.
DESSE, G. et M. H. du BUIT
  1970-1971 Diagnostic des pièces rachidiennes des Téléostéens et des Chondrichthyens.1. Gadidés, 71 pp (1970) 2. Chondrichthyens, 79 pp. (1971) *L'Expansion Scientifique. Paris.*
DESSE, G, et J. DESSE
  1976   Diagnostic des pièces rachidiennes des Téléostéens et des Chondrichthyens. (3): Téléostéens d'eau douce. 108 pp. *L'Expansion Scientifique.* Paris.
DESSE, G. et J. DESSE
  1983   L'identification des vèrtebres de Poissons: applications au matériel issu de sites archéologiques et paléontologiques. *Arch. Sc. Genève* 36 (2): 291-296
DEMOLLE, R.; H. N. MAIER and H. H. WUNDSCH
  1964   *Handbuch der Binnenfischerei Mitteleuropas. Band II. Anatomie der Fische, von Wilhem Harder.* Stutgart: E. Schweizerbartische Verlagbuchhandlung. Vol. I: 308 pp. Vol. II:96 pp. + 19 plates.
DEVILLERS, Ch.
  1958   Le crâne des Poissons. In: *Traité de Zoologie.* Edited by P.P. Grassé. Paris 13 (1): 551-687.
DOBBEN, W. H. van,
  1935   Uber den Kiefermechanismus der Knochenfische. *Arch. Neerl. Zool.* 2 (1) : 1 -72
DUMERIL,. A. M. C. Ichthyologie analytique. (Essai d'une classification naturelle des 1856 poissons). *Mém. Acad. Sci.* Vol 27
EATON, T. H. jr.
  1935   Evolution of the upper jaw mechanism in teleost fishes. *J. Morphol.* 58 : 157 - 169
EATON, T. H. jr
  1945   Skeleton supports of the median fins of fishes. *J. Morph.* 76: 193-21

EASTMAN, J. T. and J. C. UNDERHILL
    1973   Intraspecific variation in the pharyngeal tooth for males of some
            cyprinid fishes *Copeia* 1973 (1) : 45 - 53
EDGEWORTH, F. H.
    1923   On the development of the hypobranchial, branchial and pharyngeal
            muscles of *Ceratodus*. With a note on the development of the quadrate
            and epihyal. *Quart. J. Microsc. Sci.* 67 : 325 - 368
EMELIANOV, S. W.
    1973   The origin of the dorsal fins in Teleostei with references to the evolution
            of this group of fishes. *Ichthyologia.* 5 (1) : 9 - 19
EVANS, H. E. and E. E. DEUBLER
    1955   Pharyngeal Tooth Replacement in *Semotilus atromaculatus* and
            Clinostomus elongatus, Two Species of Cyprinid Fishes.
            *Copeia.*1955(1): 31-41.
FAGADE, S. O.
    1974   Age determination in *Tilapia melanotheron* (Ruppell) in the Lagos
            Lagoon,
            Lagos , Nigeria. 71-77. In: *Ageing of fish.* (Editor) T. B. Bagenal.
            Unwin Brothers Limited. Surrey, England.
FAHY, W. E.
    1972   Influence of temperature change on the number of vertebrae and caudal
            fin rays in *Fundulus majalis* (Walbaum). *J. Cons. Int. Expl. Mer.* 34
            (2): 217 -23. FAO (Food and Agriculture Organization of U.N.)
    1970   Conversion factors: North Atlantic Species. *Bull. Fish. Stat.* 25 : 1 - 71
FITCH, J. E.
    1958   Otoliths - their importance and uses. *Amer. Malacol. Union. Ann. Rep.*
            1958 , 41
FITCH, J. E.
    1975   Fish remains from a Chumash village site (Ven-87) at Ventura,
            California. In: *3500 years on one city block, Ventura Mission Plaza*
            Archaeological Project 1974. (Roberta S. Greenwood, editor). A report
            prepared for the Development Agency, City of San Buenaventura,
            California. 435 - 470
FOLLETT, W. I.
    1975   Fish remains from the west Berkely shellmound (Ca-Ala-307),
            Alameda County, California. *Contrib. Univ. Calif. Archaeol. Res.*
            *Fac.*No. 29. Appendix G: 123-129
FOLLETT, W. I.
    1980   Fish remains from the Karlo Site (Ca-Las-7), Lassen County,
            California. *J. Great Basin and California Anthropology.* 2 (1) :
            114-122
FOLLETT, W. I.
    1982   An analysis of fish remains from ten archaeological sites at Falcon Hill
            Washoe County, Nevada, with notes on fishing practices of the
            ethnographic Kuyuidikati Northen Paiute. Appendix A. In: *The*
            Archaeology of Falcon Hill, Winnemucca Lake, Washoe County,
            Nevada. (E. Hattori, editor 178 - 205. *Nevada State Mus. Athropol.*
            Papers No. 18
FORD, E.
    1937   Vertebrae variation in teleostean fishes. *J. Mar. Biol. Assoc. U.K.* 22 :
            1 - 60
FRANÇOIS, Y.
    1958   Recherches sur l'anatomie et le développement de la nageoire dorsale
            des Téléostéens. *Arch. Zool. Exp. Gen.* 97 : 1 -108

FRANKLIN, D. R. and L. L. SMITH
1960   Note on development of scale patterns in the northern pike, *Esox lucius*
        L. *Trans. Amer. Fish. Soc.* 89 : 93
GANS, C. and T. S. PARSONS
1981   *A photographic atlas of shark anatomy: the gross anatomy of Squalus
        acanthias.* Chicago: University of Chicago Press. 106 pp.
GARRAULT, H.
1936   Developpement des fibres d'élastoïdine (actinotrichia) chez les
        Salmonides.*Arch. Anat. Micr.* Paris. 32 : 105-137
GILBERT, P. W.; R. F. MATHEWSON and D. P. RALL
1967   *Sharks, skates and rays.* Baltimore: John Hopkins Press.
GOMEZ-LARRANETA, M.
1958   Sur la formule vertébrale des quelques poissons commerciaux des côtes
        de Castellón. *Rapport P-V. Réun. Cons. Int. Comm. Expl. Mer.* 14 :
        373 -377
GOLVAN, T-J
1962   *Catalogue systématique des poissons actuels.* Paris: Masson et Cie.
        277 pp.
GOODRICH, E. S.
1904   On the dorsal fin-rays of fishes living and extinct. *Q. J. Microsc. Sci.*
        47: 465-522
GOODRICH, E. S.
1906   Notes on the development, structure and origin of the median and paired
        fins of fish. *Quart. J. Microsc. Sci.* 50 : 333 - 376
GOODRICH, E. S.
1907   On the scales of fish, living and extinct, and their importance in
        classification. *Proc. Zool. Soc.* London. 1907 : 751 -754
GOODRICH, E. S.
1909   Cyclostomes and Fishes. In: *A treatise on Zoology.* Edited by
        R. Lankester. London. A. and C. Black. part IX. 1st. fascicule.
GOODRICH, E. S.
1930   *Studies on the structure and development of vertebrates.* London:
        MacMillan (Reprinted by Dover Publications, New York, 1958). Two
        volumes; 836 pp.
GOSLINE, W. A.
1960   Contribution towards a clasification of modern Isospondylous fishes.
        Bull. Br. Mus. Nat. Hist. 6(6): 327-365
GOSLINE, W.A.
1961   The perciform caudal skeleton. *Copeia.*1961 (3) : 265 - 270
GOSLINE, W. A.
1965   Teleostean phylogeny. *Copeia.* 1965 (2) : 186 - 194
GOSLINE, W. A.
1967   Reduction in branchiostegal ray number. *Copeia.* 1967 (1) : 237 - 239
GOTTBEHUT, Volkmar
1935   Otolithen und Labyrinthe verschiedener Teleostier. *Jena. Z.
        Naturwisch.* 159-196
GRAHAM, M.
1929   Studies of age-determination in fish. Part II: a survey of the literature.
        Min. Agric. Fish. Invest. Sci. Ser. 2, 11 (3) : 1 -50
GRASSE, P-P.
1958   Agnathes et Poissons. In: Traité de Zoologie. 13:1-2785 (P-P. Grassé,
        editor).
GRAY, O.
1951   An introduction to the study of the comparative anatomy of the
        labyrinth. *J. Laryngol. Otol.* 65 : 681 - 703

GREENE, C. W. & C. H. GREEN.
1914   The skeletal musculature of the king salmon. *Bull. U.S. Bur. Fish.*, 33: 21-60
GREENSPAN, R. L.
1985   Fish and fishing in northern Great Basin prehistory. MS. Ph. D. Thesis. Univ. of Oregon.
GREENWOOD, P. H.; D. E. ROSEN; S. H. WEITZMAN & G. S. MYERS.
1966   Phyletic studies of Teleostean fishes with a provisional classification of living fishes. *Bull. Am. Mus. Nat. Hist.* 131 (4) : 341-455
GREGORY, W. K.
1904   The relation of the anterior visceral arches to the chondrocranium. *Biol. Bull.* 7: 55 - 69
GREGORY, W. K.
1933   Fish skulls: a study of the evolution of natural mechanisms. *Trans. Amer. Phil. Soc.* 23 (2): 1-481
GREGORY, W. K.
1951   *Evolution emerging.* New York: Macmillan. 1013 pp.
GRAYSON, Donald K.
1984   *Quantitative Zooarchaeology. Topics in the Analysis of Archaeological Faunas.* Academic Press, Inc. 197 pp. San Diego, New York. U.S.A.
GULLAND, J. A.
1958   Age determination of cod by fin rays and otoliths. *Biarritz Symp. CNAF.* Spec. Publ. 1 : 179 - 190
GUNTHER, C. L.
1859   On sexual differences found in bones of some recent and fossil species in frogs and fishes. *Ann. Mag. Nat. Hist.* 3 ser. 3 : 377-387. 2 pl.
HAINES, R. W.
1937   The posterior end of Meckel's cartilage and related ossifications in bony fishes. *Q. J. Microsc. Sci.* 80 (1): 1-38
HAMMARBERG, F.
1937   Zur kentniss der ontogenetischen Entwicklung des Schädels von *Lepidosteus platostomus. Acta Zool. Stockholm* . 18: 209-337
HANNA, Margaret, G.
1981   An analysis of fish scales from Aschkibohkahn FbMb-1, West Central Manitoba. *Manitoba Archaeol. Quarterly.* 5 (3) : 20- 38
HARDISTY, M.W. and I. C. POTTER (editors)
1971   *Biology of lampreys.* New York. Academic Press (Two volumes)
HARRINGTON, R. W.
1955   The osteocranium of the American cyprinid fish, *Notropis bifrenatus,* with an annotated synonymy of teleost skull bones. *Copeia.* 1955 (4) : 267-290
HENNIG, W.
1950   *Grundzüge einer Theorie der phylogenetischen Systematik.* Berlin. Deutcher Zentralverlag. 370 pp.
HENNIG, W.
1966   *Phylogenetic Systematics.* Urbana: University of Illinois Press. 263 pp.
HERTWIG, O.
1874   Uber Bau und Entwicklung der Placoidschuppen und der Zähne der Selachier. *Jena Z. Med. Naturwiss.* 8 : 331 - 404
HILE, R.
1970   Body-relation and calculation of growth in fishes. *Trans. Am. Fish. Soc.* 99 (3):468-474
HOFER, H.
1945   Zür Kentniss der Suspensionsformen des Kieferbogens und deren Zusammenhänge mit den Bau des Knöcherchen Gaumens und mit der

Kinetik des Schädels bei den Knochenfischen. *Zool. J. Anat.* 69 : 321 - 404

HOLDEN, M. J.
1962   The structure of the spine of the dogfish (*Squalus acanthias*) and its use for age determination. J. Mar. Biol. Assoc. U. K. 42 (2):179 - 197

HOLLISTER, G.
1936   Caudal skeleton of Bermuda shallow water fishes. I. Isospondyli (Clupeiformes): Elopidae, Megalopidae, Albulidae, Clupeidae, Dussumieridae, Engraulidae. *Zoologica.* 21 (3) : 257-290

HOLMGREN, N. & J. RUNNSTRÖM.
1943   General morphology of the head in fish. *Acta Zool. Stockholm.* 24 (1-3): 1-188

HOLMGREN, N. & E. A. STENSIÖ.
1936   Kranium und Visceralskelett der Akranier, Cyclostomen und Fische In: *Handbuch der vergleichenden Anatomie der Wirbeltiere.* von L. Bölk. Berlin/ Wien: Urban & Schwarzenberg. Bd. 4 : 1-1016

HUBBS, C. L.
1919   A comparative study of the bones forming the opercular series of fishes. *J. Morphol.* 33 (1) : 61-71

HUBBS, C. L.
1922   Variations in number of vertebrae and other meristic characters of fishes correlated with water temperature during development. *Amer. Nat.* 56 : 360 - 372

HUBBS, C. L. and K. F. LAGLER
1958   Fishes of the Great Lakes Region. *Cranbrook Inst. Sci. Bull.* 26 : 1-213

HUREAU, J. C. and Th. MONOD (Editors)
1978   Ckeck-list of the fishes of the north-eastern Atlantic and of the Mediterranean.CLOFNAM I: 683 pp. CLOFNAM II. Supplement : 394 pp. UNESCO. Paris

HUXLEY, T. H.
1858   Theory of the vertebrate skull. *The Croonian Lecture. Proc. Roy. Soc. London.* 9 : 381-433

HUXLEY, T.H.
1859   Observations on the development of some parts of the skeleton of fishes.*Quart. J. Microsc. Sci.* 7 : 33-46

HUXLEY, T. H.
1861   Preliminary essay upon the systematic arrangement of the fishes of the Devonian epoch. *Mem. Geol. Surv. U.K.* 10 : 1-46. 2 pl. 21 figs.

JOLLIE, Malcolm
1984   Development of Cranial and Pectoral Girdle Bones of *Lepisosteus* With a Note on Scales. *Copeia* 1984 (2): 476-502

JOLLIE, Malcolm
1986   A primer of bone for the understanding of the actinopterygian head and pectoral girdle skeletons. *Can. J. Zool.* 64: 365 - 379

JORDAN, D. S.
1905   *A guide to the study of fishes.* Henry Holt and Co. New York. Vol. 1. xxvi + 624 pp. 393 figs. Vol. 2. xxii + 540 pp. 506 figs.

JORDAN, D. S.
1923   A classification of fishes. *Stanford Univ. Publ. Ser. Biol. Sci.* 3 (2): 79-243 [1963 reprint from 1923 edition]

JORDAN, D. S. and B. W. EVERMANN
1896   The fishes of North and Middle America. Part I. *U.S. Nat. Mus. Bull.* 47 (1) : 1 -1240

KAMEL, A. H.
    1952   On the origin of the dorsal and ventral ribs of bony fishes. *Bull. Zool. Soc. Egypt.* 10 : 16 21

KERR, T.
    1958   Development and structure of some actinopterygian and urodele teeth. *Proc. Zool. Soc.* London. 133 (3) : 401 - 422

KÖLLIKER, R. A.
    1858   Über die verschiedenen Typen in der mikroskopischen Struktur des Skeletts der Knochenfische. *Sitz. Phys. Med. Ges.* Würzburg. 9: 257-271

KRUMHOLZ, L. A.
    1943   A comparative study of the Weberian ossicles in North American ostariophysine fishes. *Copeia.* 1943 (1) : 33 - 40

KUSAKA, T.
    1974   *The urohyal in fishes.* Tokio. University of Tokio Press. 310 pp.

LAGLER, K. F.; J. E. BARDACH; R. R. MILLER and D. R. M. PASSINO
    1977   *Ichthyology.* 2nd edition. 506 pp. John Wiley and Sons. New York.

LE CREN, E. D.
    1951   The length-weight relationship and seasonal cycle in gonad weight. *J. Anim.Ecol.* 162 (2) : 201 - 219

LEA, EINAR
    1910   A study on the growth of herrings. *Cons. Perm. Intern. pour l'Explor. de la Mer. Public. de circonstance.* 61 : 35-57

LEE, Rosa. M. (Mrs. T. L. Williams)
    1920   A review of the methods of age and growth determination in fish by means of their scales. *Min. Agric. and Fish. Fish. Inv. Ser. II.* 4 (2): 1-32

LEIM, A. H. and W. B. SCOTT
    1966   Fishes of the Atlantic coast of Canada. *Fish. Res. Board. Canada Bull.* 155 : 1- 485

LEKANDER, B.
    1949   The sensory line system and the canal bones in the head of some ostariophysi. *Acta Zoologica.* 30 : 1-131

LEPIKSAAR, J.
    1967   The bones of birds, amphibians and fishes found at Skedemosse. In: *The Archaeology of Skedemosse.* 109-128. (Editors) Hagberg, U.E. and M. Beskow. *Royal Acad. Letters, History and Antiquities.* Stockholm.

LEPIKSAAR, J.
    1968   Skeleton collections and osteological activities at the Museum of Natural History in Gothenburg. Sweden. *Göteborgs Naturhist. Mus. Arstyck.* 15-34

LIBOIS, R. M.; C. HALLETT-LIBOIS and R. ROSOUX
    1987   Eléments pour l'identification des restes crâniens des Poissons dulçaquicoles de Belgique et du Nord de la France. 1. Anguilliformes, Gastérostéiformes, Cyprlnotormes et Perciformes. *Fiches d'Ostéologie animale pour 'Archéologie.Serie A: Poissons.* 3: 1 - 16

LINNAEUS, C.
    1758   *Systema Naturae per tria regna, secundum classes, ordines, genera, species,cum characteribus differentiis, synonymis, locis.* Ed. 10, reformata. Holmiae. L. Salvii.

LINDSEY, C. C.
    1955   Evolution of meristic relations in the dorsal and anal fins of Teleost fishes. *Trans R. Soc. Series III.* 49 (5) : 35 - 49

LINDSEY, C. C. and M. Y. ALI
   1965   The effect of alternating temperature on vertebral count in the Medaka
           (*Oryzias latipes*). *Can. J. Zool.* 43 (1) : 99 - 104
LISON, L.
   1941   Sur la structure des dents des poissons Dipneustes. La pétrodontine.
           C.R. Séances Soc. Biol. Paris. 5-6 : 431 - 433
LOWENSTEIN, O.
   1957   The acoustico-lateralis system. In: *Physiology of Fishes*. M. Brown
           (editor).Chapter 2: 155-186. Vol. 2 : 111-135
LUNDBERG, J. G.
   1975   Homologies of the upper shoulder girdle
LUNDBERG, J. G. and J. BASKIN
   1969   The caudal skeleton of catfishes, order Siluriformes. *Amer. Mus. Nov.*
           2298 : 1 - 49
MANSUETI, A. J. & J.D. HARDY, Jr.
   1967   *Development of fishes of the Chesapeake Bay Region : an atlas of egg,
           larval and juveniles stages. Part I*. College Park: Nat. Resources.
           Univ. Maryland. Port City Press. Baltimore. Maryland. 202 pp.
MARTIN, TERRANCE J.
   1982   Animal remains from the Gross Cap site: an evaluation of fish scales
           versus fish bones in assessing the species composition of an
           archaeological assemblage. *The Michigan Archaeol.* 27 (3-4) : 77 -86
MAYHEW, O.
   1924   The skull of *Lepisosteus platostomus*. *J. Morphol.* 38 (3) : 315-346
MAYR, E.
   1969   *Principles of Systematic Zoology*. McGraw-Hill Book Co. New York.
           428 pp.
McALLISTER, D. E.
   1968   Evolution of Branchiostegals and Classification of Teleostome Fishes.
           National Museum of Canada. Bull. 221: 239 pp.
MECKEL, J. F.
   1820   *Handbuch der menschlichen Anatomie*. Halle and Berlin. 4.
MERRIMAN, Daniel
   1940   The osteology of the striped basss (*Roccus saxatilis*). *Ann. Mag. Nat.
           Hist.* 5: 55-64
MILES, A. E. W. (editor)
   1967   *Structural and chemical organization of the teeth*. Academic Press. Vol.
           I and II. 489 pp.
MONASTYRSKY, G. N.
   1930   Uber Methoden zur bestimmung des linearen Wachstums des Fische
           nach Schuppe. *Rept. Sci. Inst. Fish. Cult.* 5 (4) : 5-44
MONOD, T.
   1968   Le complexe urophore des poissons téléostéens. *Mém. Int. Afrique
           Noire.* 81: 1-705
MOOKERJEE, H. K.
   1936   The development of the vertebral column and its bearing on the study of
           evolution. *23rd Ind. Sci. Cong. Assoc. Proc.* 1936: 307 - 343
MORALES, A. and K. ROSELUND
   1979   Fish bone measurements: An attempt to standardize the measuring of
           fish bones from archaeological sites. *Steenstrupsia.* 1 - 48
MOSS, M.L.
   1972   The vertebrate dermis and the integumentary skeleton. *Amer. Zool.* 12
           (1) : 27 - 34
MOSS, M. L.
   1977   Skeletal tissues in sharks. *Amer. Zool.* 17 : 335 - 342

MOY-THOMAS, J. A.
  1938   The problem of the evolution of dermal bones in fishes. In: *Evolution.*
          Clarendon Press. Oxford. 305-31
MOYLE, Peter B. and J. J. CECH, jr.
  1982   *Fishes: and introduction to Ichthyology.* Prentice-Hall, Inc.Englewood
          Cliffs. New Jersey.
MUJIB, K. A.
  1967   The cranial osteology of Gadidae. *J. Fish. Res. Board. Canada.*24 (6) :
          1315 - 1375
MÜLLER, A.,
  1853   Beobachtung zur vergleichende Anatomie der Wirbelsäule. *Arch, Anat.
          Physiol.* 1853: 260-316
MÜLLER, J.
  1844   Uber den bau und die frenzen der Gadoiden und über des naturliche
          System der Fische. *Abh. Akad. Wiss. Berlin.* 177-216
MUNDELL, Raymond, L
  1975   An illustrated Osteology of the Channel catfish (*Ictalurus punctatus*).
          Midwest archaeological Center. Occas. Studies in Anthrop. 2 : 1-1
MUUS, B. J.
  1971   *Freshwater fishes of Britain and Europe.* London: Collins. 222 pp.
MYERS, G. S.
  1958   Trends in the evolution of teleostean fishes. *Stanford Ichthyol. Bull.* 7
          (3) : 27-30
NELSON, E. M.
  1949   The opercular series of the Catostomidae. *J. Morphol.* 559-567
NELSON, G. J.
  1968   Gill arches of teleostean fishes of the division Osteoglossomorpha. *J.
          Linn. Soc. London.* 47 (312) : 261-278
NELSON, G. J.
  1969   Gill arches and the phylogeny of fishes, with notes on the classification
          of vertebrates. *Bull. Am. Mus. Nat. Hist.* 141 (4) : 479-552
NELSON, J. S.
  1971   Absence of the pelvic complex in ninespine stickleback, *Pungitius
          pungitius,* collected in Ireland and Wood Buffalo National Park region,
          Canada, with notes on meristic variation. *Copeia.* 1971 (4) : 707-717
NELSON, J. S.
  1976   *Fishes of the World.* New York: John Wiley and Sons
NICHY, F. E.
  1974   Bibliography of age validation studies. *Int. Comm. Northwest. Atl.
          Fish. Res.,* Doc. 74/110 : 1-3
NIEHOFF, A.
  1952   Otoliths as ornaments. *Wisc. Archaeol.* 33 : 223 - 224
NORMAN, J. R.
  1958   *A history of fishes.* 5th edition. London. 463 pp.
NURSALL, J. R.
  1963   The hypurapophysis, an important element of the caudal skeleton.
          Copeia. 1963 (2) : 458-459
NYBELIN, O. von,
  1963   Zur Morphologie und Terminologie des Schwanzeskelettes der
          Actinopterygier. *Ark. Zool.* 15 (2) : 485-516
NYBELIN, O. von.,
  1973   Comments on the caudal skeleton of Actinopterygians.
          In : *Interrelationships of Fishes.* P.H. Greenwood, R. S. MIles, and
          C. Patterson. London: Academic Press. 369-372. [Suppl. 1 to the
          Zool. J. Linn. Soc. London]

OLSEN, Stanley J.
    1968   Fish, amphibian and reptile remains from archaeological sites. Part I.
    Southeastern and Southwestern United States. *Paps. Peabody Mus.
    Archaeol. Ethnol.* Harvard Univ. 56 (2): i-xv; 1-94

OLSEN, Stanley, J.
    1971   Zooarchaeology: Animal bones in Archaeology and their interpretation.

OWEN, R.
    1848   Report of the archetype and homologies of the vertebrate skeleton. *Br.
    Assoc. Adv. Sci. Rep.* 16th Meeting: 169-340

OWEN, R.
    1866   *On the anatomy of vertebrates.* London: Logmanus, Green. 3 vol.

PANELLA, G.
    1971   Fish otoliths: daily growth layers and periodical patterns. *Science.* 173:
    1124-1127

PARKER, W. K.
    1868   A monograph on the Structure and Development of the Shoulder-girdle
    and Sternum in the Vertebrata. London

PARKER, W. K.
    1874   On the structure and development of the skull in the salmon (*Salmo salar*
    L) *Phil. Trans. R. Soc.* London. Part I. 163 : 95-145

PARR, A. E
    1930   Jugostegalia: an accessory skeleton in the gill cover of the eels of the
    genus *Myrophis. Copeia.* 1930 (3) : 71-73

PARRINGTON, F. R.
    1967   The identifiation of the dermal bones of the head. *J. Linn. Soc. (Zool)
    London.* 47 (311) : 231- 239

PATTERSON, C.
    1968   The caudal skeleton in lower Liassic pholidophorid fishes. *Bull. Br.
    Mus. Nat. Hist.* 16 (5): 203- 239

PATTERSON, C.
    1973   Interrelationships of Holosteans. In: *Interrelatioships of Fishes.* Edited
    by P.H. Greenwood, R.L. Miles and C. Patterson. London: Academic
    Press. 233-305

PATTERSON, C.
    1975   The braincase of Pholidophorid and Leptolepid fishes, with a review of
    the Actinopterygian braincase. *Phil. Trans. Royal Soc.* Series B. 269:
    275-579. London.

PEHRSON, T.
    1944   Some observations on the development and morphology of the dermal
    bones in the skull of *Acipenser* and *Polyodon. Acta Zool. Stockholm.*
    25: 27-48

PENCSAK, T.
    1965   Morphological variation of the stickleback (*Gasterosteus aculeatus*) in
    Poland. *Zool. Polon.* 15 : 3-49

PEYER, B.
    1968   *Comparative odontology.* Chicago: Univesity of Chicago Press.
    347 pp.

PHILLIPS, J. B.
    1948   Comparison of calculated fish lengths based on scales from different
    body areas of the sardine, *Sardinops caerulea. Copeia.* 1948 (2) :
    99 - 116

PRADHAN, L. B. and P. N. KAPADIA
    otolith and bone, and rate of ossification. *J. Univ. Bombay.* 22 (3) :
    32 - 40

PRIEGEL, G. R.
1963   Use of otoliths to determine length and weight of ancient freshwater
        drum in the Lake Winnebago Area. *Wisconsin Acad. Sci. Arts and*
        Letters. 52 : 27 - 35
PRINCE, E. E.
1893   The development of the pharyngeal teeth in the Labridae. *Br. Assoc.*
        Adv. Sci. Rep. 773 pp.
RAMASWAMI, L. S.
1948   The homalopterid skull. *Proc. Zool. Soc. London.* 118 : 515-538
RASS, T. S. & U. LINDBERG.
1971   Modern concepts of the classification of living fishes. *J. Ichthyol.* 11
        (2): 302-319
REED, Ch. A.
1963   Osteo-Archaeology. 204-216. In: *Science in Archaeology* (editors)
        D. Brothwell, E.Higgs and G. Clark. Basic Books Inc. Publishers.
        New York
REGAN, C. T.
1909.   The classification of the Teleostean fishes. *Am. Mus. Nat. Hist.* 3:
        75-86
REGAN, C. T.
1910   The caudal fin of the Elopidae and some other Teleostean fishes. *Ann.*
        Mag.Nat. Hist. 5 (8)
REGAN, C. T.
1916   The morphology of the Cyprinodont fishes of the subfamily
        Phallostethinae. *Proc. Zool. Soc. London.* 1-26
REGAN, C.T.
1929   Fishes. In: *Encyclopedia Britannica: a new survey of universal*
        knowledge. 14th edition. London: Encycl. Britann. Vol. 9 : 305-328
REGIER, H. A.
1962   Validation of the scale method for estimating age and growth of
        bluegills.*Trans. Am. Fish. Soc.* 91(4) : 362-373
REIF, W. E.
1978   Shark dentitions: morphometric processes and evolution. *Neues Jahrb.*
        Geol. Paläeontol. Abh. 157 (1-2) : 107 - 116
RIDEWOOD, W. G.
1904   On the cranial osteology of the fishes of the families Elopidae and
        Albulidae.*Proc. Zool. Soc. London.* 2 : 35-81
RIDEWOOD, W. G.
1921   On the calcification of the vertebral centra in sharks and rays. *Phil.*
        Trans. R. Soc. London. Ser. B. 210 : 311 - 407
ROJO, Alfonso, L.
1959   Sampling data for Spain. In: *Sampling Year Book for the year 1957.*
        Vol. 2: 112 -119. ICNAF. Halifax. N. S. Canada.
ROJO, Alfonso, L.
1976   Osteología de la merluza argentina (*Merluccius hubbsi.* Marini 1933).
        Bol. Inst. Esp. Oceanogr. 219 : 1-61
ROJO, Alfonso, L.
1977   El crecimiento relativo del otolito como criterio identificador de
        poblaciones del bacalao del Atlántico noroeste. *Inv. Pesq.* 41 (2) :
        239-261
ROJO,Alfonso, L.
1986   Live length and weight of cod (*Gadus morhua*) estimated from various
        skeletal elements. *North American Archaeologist.* 7 (4): 329-351
ROJO, Alfonso, L.
1987   Excavated fish vertebrae as predictors in bioarcheological research.

North American Archaeologist. 8 (3) : 209- 225

ROJO, Alfonso, L
1988   Diccionario enciclopédico multilingüe de anatomía de peces. *Instit. Esp.
Oceanogr.* Monografía no. 3. 564 pp. Madrid

ROJO. Alfonso, L.
1989   X-rays as a tool for distinguishing the vertebrae of cod (*Gadus
morhua*), saithe (*Pollachius virens*), cusk (*Brosme brosme)*, and
haddock (*Melanogrammus aeglefinus*) of the fish family Gadidae.
(In press) BAR. International Series.

ROJO, Alfonso, L. and D. A. A. CAPEZZANI
1971   Características morfométricas y merísticas de la merluza argentina
(*Merluccius merluccius hubbsi*). *Inv. Pesq.* 35 (2) : 589-637

ROJO, Alfonso, L. and P. RAMOS
1983   Tiempo y orden de aparición de las escamas en el salmón del Atlántico
(*Salmo salar*). *Doñana Acta Vertebrata.* 10 (1) : 5 - 17

ROMER, A. S.
1933   *The Vertebrate Body.* The University of Chicago Press, 473 pp.

ROMER, A. S. and T.S. PARSONS
1978   *The Vertebrate Body.* W.B. Saunders Co. Philadelphia. London.
Toronto.

ROSEN, D. E. and P. H. GREENWOOD
1970   Origin of the Weberian apparatus and the relationships of the
ostariophysan and gonorynchiform fishes. *Amer. Mus. Nov.* 2428 :
1 - 25

ROSENLUND, Knud
1986   The sting ray, *Dasyatis pastinaca.* (L.) in Denmark. 123-128 In: *Fish
and Achaeology. Studies in osteometry, taphonomy, seasonality and
fishing methods.* (Editors) D.C. Brinkhuisen and A. T. Clason. BAR
International Series.

ROSTLUND, Erhard
1952   Freshwater fish and fishing in native North America. *Univ. Cal. Publ.
Geography.* 9 : 1 - 313.

ROYCE, W. F.
1942   Standard length versus total length. *Trans. Am. Fish. Soc.* 71:270 - 274

RYDER, M. L.
1976   Remains of fishes and other aquatic animals. 204-312. In: *Science in
Archaeology.* (Editors). Brothwell, D., E. Higgs and G. Clark. Basic
Books Inc. Publishers. New York.

SAGEMEHL, M.
1885   Beiträge zur vergleichenden Anatomie der Fische. III. Das Cranium der
Chariciniden nebst allgemeinen Bemerkungen über die mit einem
Weberschen Apparat versehenen Physostomenfamilien. *Morphol.
Jahrb.* 10: 1-119.

SAGEMEHL, M.
1891   Beiträge zur vergleichenden Anatomie der Fische. IV. Das Cranium der
Cyprinoiden. *Morphol. Jahrb.* 17 : 489-595.

SCHAEFFER, B.
1967   Osteichthyan vertebrae. *J. Linn. Soc. London.* 47 (311) : 185-195

SCHAEFFER, B. & D. E. ROSEN.
1961   Major adaptive levels in the evolution of the actinopterygian feeding
mechanism. *Am. Zool.* 1 : 187-204

SCHMALHAUSEN, I. I.
1923   Der suspensorialapparat der Fische und das Problem de
Gehörknöchelchen (Vorläufige Mitteilung). *Anat. Anz.* 56 : 534 - 543

SCHUCK. A.
  1949   Problems in calculating size of fish at various ages from proportional
          measurements of fish and scale sizes. *J. Wildl. Manag.* 13 (3) :
          298 - 303
SCOTT, J. S.
  1977   Back-calculating fish lengths and Hg and Zn levels from recent and
          100-year-old cleithrum bones from Atlantic cod (*Gadus morhua*). *J.
          Fish. Research Board. Canada.* 34 : 147-150
SCOTT, T.
  1906   Observations on the otoliths of some teleostean fishes. *Glasgow Rept.
          Fish. Board.* 24 : 48-82
SCOTT, W. B. and E. J. CROSSMAN
  1973   Freshwater fishes of Canada. *Fish. Res. Board Canada. Bull.* 184 :
          1 - 966
SEGERSTRALE, C.
  1933   Uber scalimetrische Methoden zur Bestimmung des linearen Wachstum
          bei Fischen, insbesondere bei *Leuciscus idus* L. *Abramis brama* L. und
          Perca fluviatilis L. *Acta Zool. Fenn.* 15 : 1-168.
SEWERTZOFF, A. N.
  1926   Studies on the bony skull of fishes. I: Structure and development of the
          bony skull of *Acipenser ruthenus*. II: Primary structure of the bony
          skull of fishes. *Q. J. Microsc.Sci.* 70 : 451-540
SEWERTZOFF, A. N.
  1928   The head skeleton and muscle of *Acipenser ruthenus*. *Acta Zool. Stock.*
          9: 193-319
SEYMOUR, A.
  1959   Effect of temperature upon the formation of vertebrae and fin rays in
          young chinook salmon. *Trans. Amer. Fish. Soc.* 88 (1) : 58 - 69
SHEPHERD, C. E.
  1910   The "asteriscus" in fishes. *The Zoologist.* 4 series (14) : 57 - 62
SHERRIFF, Catherine
  1922   Herring investigations. Report on the mathematical analysis of random
          samples of herrings. *Fish. Bd. Scotland, Sci. Inv.* no. 1: 1-22
SHUTE, C. C. D.
  1972   The composition of vertebrae and the occipital region of the skull.
          In: *Studies in vertebrate evolution.* 21 - 34. K. A. Joysey and T. S.
          Kemp (editors). Edinburg: Oliver and Boyd.
SIMPSON, G. G.
  1961   *Principles of animal taxonomy.* New York: Columbia University Press.
          247 PP.
SMITH, H. M.
  1960   *Evolution of Chordate Structure.* Holt, Reinhart and Wiston. Inc.
          529 pp.
SMITH, S. M.
  1962   Classification of structural and functional similarities in biology. *Syst.
          Zool.* 11: 45-47
SPRINGER, V. G. and J. A. F. GARRICK
  1964   A survey of vertebral numbers in sharks. *Proc. U.S. Mus.* 116 (3496) :
          73 -96
STARKS, E. C.
  1901   Synonomy of the fish skeleton. *Proc. Washington Acad. Sci.* 3 :
          507-539
STARKS, E. C.
  1926   Bones of the ethmoid region of the fish skull. *Stanford Univ. Publ.
          Univ. Ser. Biol. Sci.* 4 (3) : 139-338

STENSIÖ, E. A.
1947   The sensory lines and dermal bones of the cheek in fishes and
        amphibians. Stockholm: Almqvust & Wiksells. [*K. Sven.
        Vetenskapsakad. Handl.* Ser. 3, Bd. 24 (3)]
STÖHR, P.
1882   *Zur Entwicklungsgeschichte des Kopfskelets der Teleostier.* Leipzig:
        F.C.W. Vogel [Fests. z. Feier d. 300 jähr Best. d. Julius-Max-Univ.
        z. Würzburg]
STOKLEY, P. S.
1952   The vertebral axis of two species of centrarchid fishes. *Copeia.* 1952
        (4) : 255-261
SUMMERFELT, R. C. and G. E. HALL
1987   *Age and growth of fish.* Iowa State University Press. Ames. 544 pp.
SWINNERTON, H. H.
1902   A contribution to the morphology of the teleostean head skeleton based
        upon a study of the development of the three-spined stickleback
        (*Gasterosteus aculeatus*) *Q. J. Microsc. Sci.* 45 : 503-593
TATARA, K.; Y. YAMAGUCHI and T. HAYASHI
1962   Study for the restoration of length and weight of prey fishes, found in
        the stomachs of predators, by graphic representation using column
        length of fish vertebrae. *Bull. Naikai Region. Fish. Res. Lab. Fish.
        Agency.* 16 : 199 - 228
TCHERNAVIN, V.V.
1938   Notes on the chondrocranium and branchial skeleton of *Salmo. Proc.
        Zool. Soc. London.* 108 B : 347-364
TEMPLEMAN, W. and H.J. SQUIRES
1956   Relationship of otolith lengths and weights in the haddock
        Melanogrammus aeglefinus (L) to the rate of growth of the fish. *J. Fish.
        Res. Board. Canada.* 13 (4) : 467- 487
THILO, O.
1908   Die Bedeutung der Weberschen Knöchelchen. *Zool. Anz.* 32 : 777-789
THOMPSON, J. R. and S. SPRINGER
1965   Sharks, skates, rays and chimaeras. *U. S. Fish Wildl. Serv. Circ.* 228 :
        1 - 18
VANDEWALLE, Pierre
1975   Contribution à l'Etude anatomique and fonctionelle de la Région
        céphalique de *Gobio gobio* (L.) (Pisces, Cyprinidae). *Forma et Functio.*
        8:331-360
VAN NEER, Wim
1989   Contribution à l'ostéométrie de la perche du Nile *Lates niloticus*
        (Linnaeus, 1758). *Fiches d'Ostéologie animale pour l'Archéologie.
        Série: Poissons.* Centre de Recherches Archéologiques du CNRS. 1-9
        pp. 58 diagrammes
VERAN, Monette
1989   Les éléments accesoires de l'arc hyoïdien des poissons téléostomes
        (Acanthodiens et Osteichthyens) fossiles et actuels. Mémoirs du
        Muséum d'Histoire Naturelle. Série C. Sciences de la Terre. 54: 13-98.
        VII planches.
VLADYKOV, V. D.
1934   Geographical variation in the number of rows of pharyngeal teeth in
        cyprinid genera. *Copeia.* 1934 (3) : 134 - 136
VROLIK, A. J.
1873   Studien über die Verknocherung und die Knochen der Schädels der
        Teleostei. *Niederlandisches Arch. Zool.* 1(3) : 219-318

WARNER, K. & K. HARVEY.
1961   Body-scale relationship in landlocked salmon, *Salmo salar. Trans. Am. Fish. Soc.* 90 : 457-461

WEBER, E. H.
1820   *De aure et auditu hominis et animalium. I. De aure animalium aquatilium.* Leipzig: Gerhard Fleischer

WEISEL, G. F.
1967   The Pharyngeal Teeth of larval and juvenile Suckers (*Catostomus*). *Copeia.*1967. (1) : 50-54

WEITZMAN, S. H.
1962   The osteology of *Brycon meeki,* a generalized characid fish, with an osteological definition of the family. *Stanford Ichthyolo.* Bull. 8 (1) : 3-77

WEITZMAN, S. H.
1967   The origin of the Stomiatoid Fishes with comments on the classification of Salmoniform fishes. *Copeia.* 1967 (3) : 507-540

WESTOLL, T. S.,
1936   On the structures of the dermal ethmoid shield of *Osteolepis. Geol.Mag.* 73: 157-171

WHITE, E. C.
1937   Interrelationship of the Elasmobranchs, with a key to the order Galea. *Bull. Amer. Mus. Nat. Hist.* 74 : 25 - 138

WHITEHOUSE, R. H.
1910   The caudal fin of the Teleostomi. *Proc. Zool. Soc.* London. 589-1033

WIJHE, J. W. van
1882   Ueber das Visceralskelett und die Nerven des Kopfes bei Ganoiden und von *Ceratodus. Niederländisches Arch. Zool.* 5 : 207-320, pls. 15-16

WIJHE, J. W. van
1922   Frühe Entwicklungsstadien des Kopfes und Rumpfskelett von *Acanthias vulgaris. Brijdr. dierkd.* 22: 271-279

WILLIAMSON, W. C.,
1849   On the microscopic structures of the scales and dermal teeth of some Ganoid and Placoid fishes. *Phil. Trans. R. Soc.* London. 111: 435-475

WILLUGHBY, F.
1686   *De historia piscium.* (1978 Reprint by Arno Press Inc. New York)

WINTEMBERG, W. J.
1919   Archaeology as an aid to Zoology. *The Canadian Field-Culturist.* 33 (4): 63-66

WITT, A.
1960   Length and weight of ancient freshwater drum, *Aplodinotus grunniens,* calculated from otoliths found in Indian middens. *Copeia.* 1960 (3): 181- 185

YERKES, R. W.
1977   An analysis of the fish bone and scale remains from the Larson Site (Fv1109): A Spoon River variant Mississippian town in the central Illinois Valley. *Master Thesis (Anthrop.)* Univ. Wisconsin-Madison

8.   GLOSSARIES

A.   ENGLISH   GLOSSARY

This glossary contains the words not defined in the descriptive section but referred to under the corresponding number.

B. FRENCH   GLOSSARY

On a énuméré ici les mots français correspondants aux termes anglais définis
dans la Section Descriptive

2	Acanthoptérygiens	66	cartilage cératohyal
3	acanthotriches	80	cartilage coracoïde
5	Actinoptérygiens	207	cartilage de Meckel
6	actinostes	184	cartilage ischio-pubien
7	actinotriches	318	cartilage scapulaire
9	adnasal	263	ceinture pelvienne
10	Agnathes	261	ceinture scapulaire
12	alisphénoïde	65	cératobranchiaux
16	angulaire	66	ceratohyal
17	antorbitaire	213	charactéristiques méristiques
19	apophyse	224	charactéristiques
374	appareil de Weber		morphométriques
154	arc hémal	68	chevrons
166	arc hyoïdien	69	Chondrichthyens
203	arc mandibulaire	70	Chondrocrâne
231	arc neural	71	Chondrostéens
21	articulaire	73	circumorbitaires
23	articulation	74	classification des Poissons
26	astériscus	75	clavicule
28	atlas	76	cléithrum
30	autopalatin	369	colonne vertébrale
31	autoptérotique	77	condyle
32	autosphénotique	334	console oculaire
34	axonoste	78	convergence
35	azygoste	79	coracoïde
184	barre ischio-pubienne	82	coronoïdes
36	basapophyse	83	coronomeckélien
37	baséoste	310	côtes
38	basibranchiaux	85	crâne
39	basicrâne	327	crâne
40	basihyal	272	crâne platybasique
41	basioccipital	360	crâne tropybasique
42	basiptérygie	86	crête
43	basisphénoïde	236	crête occipitale
44	bélophragme	87	Crossoptérygiens
321	bouclier	93	Cyclostomes
48	Brachioptérygiens	96	dentaire
50	branchicténies	99	denticules cutanés
49	branchie	353	dents
145	branchies	4	dents acrodontes
51	branchiocrâne	56	dents caniniformes
52	branchioperculaire	164	dents cornées
144	branchiospines	58	dents en cardes
155	canal hémal	259	dents en pavé
232	canal neural	371	dents en soie
322	canaux semi-circulaires	58	dents en velours
29	capsule auditive	158	dents hétérodontes
302	carré	162	dents homodontes
59	cartilage	175	dents incisiformes

C. GERMAN GLOSSARY

Dieses alphabetische geordnete Wörterverzeichnis entspricht den englischen
Wörtern, die im beschreibenden Abschnitt definiert sind

2	Acanthopterygier	53	Branchiostegalstrahlen
3	Acanthotrichen	260	Brustflossen
33	Achsenskelett	56	caniniforme Zähne
4	acrodonte Zähne	58	cardiforme Zähne
5	Actinopterygier	65	Ceratobranchialia
7	Actinotrichen	66	Ceratohyale
9	Adnasale	66	Ceratohyalknorpel
14	Afterflosse	67	Ceratotrichen
3	Akantotrichen	71	Chondrosteer
4	akrodonte Zähne	235	Chorda
5	Aktinopterygier	72	Cilienschuppen
7	Aktinotrichen	73	Circumorbitalia
11	Alarschuppen	73	Circumorbitalknochen
12	Alisphenoid	76	Cleithrum
13	amphicoeler Wirbel	77	Condylus
13	amphizoeler Wirbel	79	Coracoid
14	Analflosse	80	Coracoidknorpel
16	Angulare	82	Coronoidea
370	Ansicht	81	Coronoidfortsatz
17	Antorbitale	84	Cosmoidschuppen
19	Apophyse	85	Cranium
21	Articulare	87	Crossopterygier
22	Articularfortsatz	88	Ctenien
25	aspondyler Wirbel	89	Ctenoidfische
26	Asteriscus	90	Ctenoidschuppen
27	asterospondyler Wirbel	91	Cycloidfische
28	Atlas	93	cyclospondyler Wirbel
345	Aufhängeapparat	94	Cyclostomen
24	aufsteigender	92	Cykloidschuppen
128	Aussenskelett	209	Deckknochen
19	Auswuchs	61	Deckknochen der
30	Autopalatinum		Chorda
31	Autopteroticum	96	Dentale
32	Autosphenoticum	99	dermale Dentikel
36	Basapophyse	100	Dermarticulare
38	Basibranchialia	102	Dermatocranium
39	Basicranium	102	Dermatokranium
40	Basihyale	107	Dermatotrichen
41	Basioccipitale	103	Dermopalatinum
42	Basipterygium	105	Dermosphenoticum
43	Basisphenoid	106	Dermosupraoccipitale
262	Bauchflossen	108	diplospondyler Wirbel
263	Beckengürtel	109	Dipnoer
45	beryciformes Loch	118	distales Segment der
322	Bogengänge		Flossenträger
48	Brachiopterygier	37	distales Segment der
50	Branchictenien		Flossenträger
51	Branchiocranium	329	Dornen
52	Branchioperculum	110	Dorsalflosse

D. LATIN GLOSSARY

47	labyrinthus osseus	167	os hyomandibularis
192	lagena	170	os hypohyale
193	lapillus	179	os innominatum
198	lepidotrichia	180	os intercalare
196	linea lateralis	181	os interhyale
152	longitudo capitis	183	os interoperculare
359	longitudo maxima	183	os interoperculum
202	Malacopterygii	186	os jugale
206	maxilla	190	os lacrimale
239	membrana branchialis	191	os lacrimojugale
217	mesopterygium	200	os linguale plattum
219	metapterygium	206	os maxillare
227	myxopterygium	212	os mentomeckelium
234	neurocranium	214	os mesethmoideum
370	norma	215	os mesocoracoideum
110	notopterygium	218	os mesopterygoideum
260	omopterygium	220	os metapterygoideum
46	os	228	os nasale
9	os adnasale	238	os operculare
12	os alisphenoidale	238	os operculum
16	os angulare	241	os opisthoticum
17	os antorbitale	243	os orbitosphenoideum
21	os articulare	250	os palatinum
30	os autopalatinum	255	os parasphenoideum
31	os autopteroticum	257	os parietale
32	os autosphenoticum	258	os parietooccipitale
40	os basihyale	201	os pharyngeum inferior
41	os basioccipitale	274	os pleurosphenoideum
42	os basipterygium	276	os postcleithrum
43	os basisphenoideum	279	os posttemporale
52	os branchioperculum	281	os præarticulare
66	os ceratohyale	282	os præethmoideum
76	os cleithrum	283	os præfrontale
79	os coracoideum	284	os præmaxilla
83	os coronomeckelium	286	os prænasale
179	os coxæ	286	os præoperculare
96	os dentale	287	os præoperculum
100	os dermoarticulare	288	os præorbitale
103	os dermopalatinum	290	os prævomere
105	os dermosphenoticum	294	os proethmoideum
106	os dermosupraoccipitale	295	os prooticum
113	os ectocoracoideum	297	os pterosphenoideum
114	os ectopterygoideum	298	os pteroticum
116	os entopterygoideum	301	os pterygoideum
122	os epihyale	303	os quadrato-jugale
123	os epioccipitale	302	os quadratum
124	os epioticum	34	os radiiferum
126	os ethmoidale	308	os retroarticulare
126	os ethmoideum	311	os rostrale
194	os ethmoideum laterale	317	os scapulum
127	os exoccipitale	323	os septomaxillare
140	os frontale	328	os sphenoticum
146	os glossohyale	332	os squamosum
150	os gulare	335	os suboperculare
195	os gulare laterale	335	os suboperculum

E. RUSSIAN GLOSSARY

Этот словник включает в себя слоба, которие
соответствуют английским словам, опредслённым
в описательном разделе

E. SPANISH GLOSSARY

Este glosario consta de las palabras correspondientes a los términos ingleses
definidos en la Section Descriptiva

10. LATIN LIST

The following list includes the Latin terms used in some works
to describe certain anatomical structures and the relative position of the bones.
(Abbr: *adj*=adjective; *f*=feminine; *m*=masculine; *n*=neuter; *pl*; plural) [1]

---

aboralis, -e *adj*	aboral
angulus (pl: anguli) *m*	angle
anterioris, -e *adj*	anterior, foremost
apertura (pl: aperturæ) *f.*	opening
apex (pl: apices) *m*	top, summit,
ascendens (pl: ascendentes) *adj*	ascending
auditivus, -a, -um *adj*	auditive
capitulum (pl: capitula) *n*	small head
caudalis, -e *adj*	caudal, related to the tail
chondrogenus, -a, -um *adj*	chondral or cartilaginous in origin
collum (pl: colla) *n*	neck; narrowing
collus (pl: colli) *m*	neck; narrowing
corpus (pl: corpora) *n*	body
cranialis, -e *adj*	craneal
crista (pl: cristæ) *f*	crest
crus (pl: crura) *n*	foot, leg, shin
descendens (pl: descendentes) *adj*	descending
desmogenus,-a,-um *adj*	membraneous in origin
dexter, dextera, dexterum *adj*	right
dexter, dextra, dextrum *adj*	right
dorsalis, -e *adj*	dorsal
epaxonicus, -a, -um *adj*	epaxial; above the body axis
externus, -a, -um *adj*	external
facies (pl: facies) *f*	external form, appearance
hypaxonius,-a, -um *adj*	hypaxial; below the body axis
impressio (pl: impressiones) *f*	impression, mark
internus, -a, -um *adj*	internal
lateralis,-e *adj*	lateral
limbus (pl: limbi) *m*	limb
margo (pl: marginis) *f.*	margin, border, hem
medialis, -e *adj*	medial
musculus (pl: musculi) *m*	muscle
nasalis, -e (pl: nasales) *adj*	nasal
norma (pl: normæ) *f*	view, face
oralis, -e *adj*	oral
os (pl: ossa) *n*	bone
ossiculum (pl. ossicula) *n*	small bone
oticalis, -e *adj*	otic
pars (pl: partes) *f*	part, section
planum (pl: plana) *n*	plane, surface
ruga (pl: rugæ) *m*	crease, wrinkle
sectio (pl: sectiones) *f*	section, part
septum (pl: septa) *n*	wall
serratus, -a, -um *adj.*	serrated, toothed
sinister, sinistra, sinistrum *adj*	left
sive *conjunction*	or

stria (pl: striæ) *f.*	furrow, channel
sulcus (pl: sulci) *m*	deep furrow
superficies (pl: superficies) *m*	surface
symphysicus, -a, -um *adj*	symphyseal
tuberositas (pl: tuberositates) *m*	tuberosity
ventralis, -e *adj*	ventral

[1] Adjectives with two endings: -<u>is</u> for the masculine and feminine and -<u>e</u> for the neuter, form the plural in <u>es</u> and <u>ia</u>, respectively. As adjectives, they should agree with the noun in number, gender, and case.

Examples: sing: processus caudalis; facies aboralis; corpus laterale.
pl: processus caudales; facies aborales; corpora lateralia

Adjectives with three endings : -<u>us</u> for the masculine, -<u>a</u>, for the feminine, and -<u>um</u> for the neuter, form the plural in -<u>i</u>, -<u>æ</u>, and -<u>a</u>, respectively. They should agree with the noun in gender, number, and case.

Examples: sing.   musculus hypaxonicus; facies symphysica; clavicula dextra; cleithrum dextrum
pl: musculi hypaxonici; facies symphysicae; claviculae dextrae; cleithra dextra

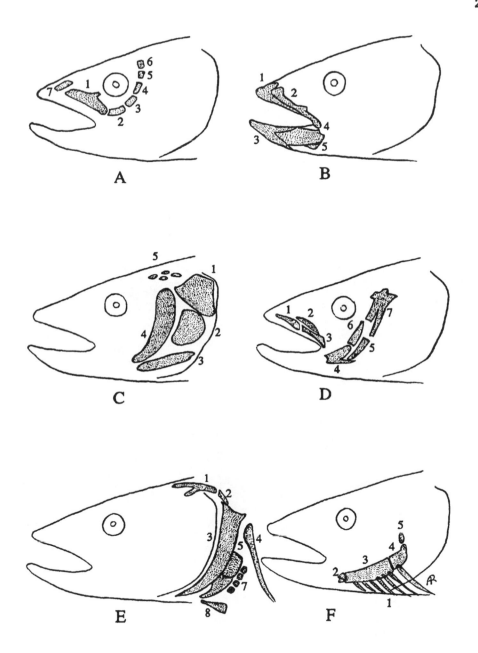

Fig. 1. Recommended stages in the dissection of the skeleton of cod (*Gadus morhua*).
[In the following illustrations, the scientific names of fishes already identified are eliminated].

Fig. 2. Dorsal view of the bones of the disarticulated braincase (*sensu lato*) of a codfish 850 mm in total length. 1. Ethmoid. 2. Lateral ethmoid. 3. Frontal (both frontals fused in cod). 4. Pterosphenoid. 5. Prootic. 6. Parietal (the right one in ventral view). 7. Sphenotic (the right one in ventral view). 8. Intercalar. 9. Pterotic. 10. Epiotic. 11. Supraoccipital. 12. Exoccipital. 13. Basioccipital. The prevomer and the parasphenoid are not shown. (See figures 20 and 21, respectively). Photo L. Niven.

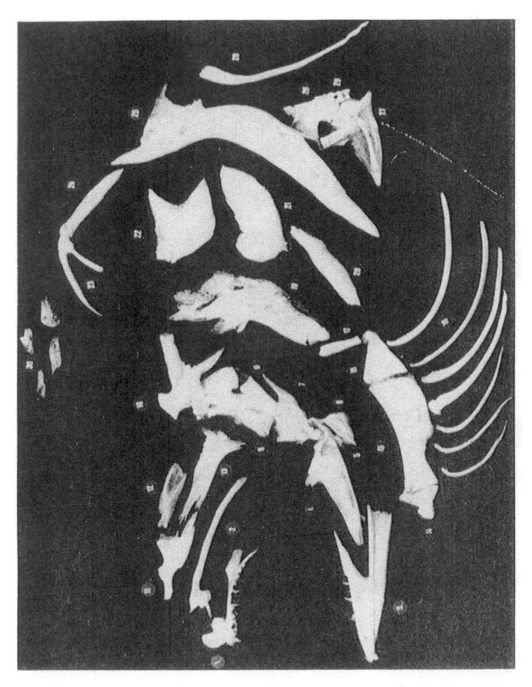

Fig. 3. Bones of the left side of the skull of the same codfish. 1. Premaxilla. 2. Maxilla. 3. Dentary. 4 and 5. Angular. 6. Retroarticular. 7. Quadrate. 8. Metapterygoid. 9. Symplectic. 10. Hyomandibula. 11. Palatine. 12. Endopterygoid. 13. Ectopterygoid. 14. Hypohyals (= dorso- and ventrohyal). 15. Ceratohyal (= anterohyal). 16. Epihyal (= posterohyal). 17. Interhyal. 18. Branchiostegal rays. 19. Preopercle. 20. Interopercle. 21. Subopercle. 22. Opercle. 23. Posttemporal. 24. Supracleithrum. 25. Cleithrum. 26. Scapula. 27. Coracoid. 28. Postcleithrum (outer view). 29. Radials. 30. Tabulars. Photo L. Niven.

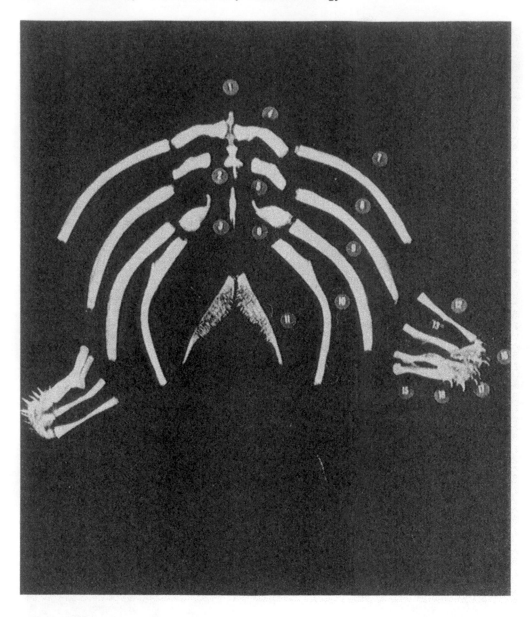

Fig. 4. Dorsal view of the bones of the gill apparatus of the same codfish. 1. Basihyal. 2. -3. Basibranchials. 4-6. Hypobranchials. 7-11. Ceratobranchials. 12-15. Epibranchials. 16-18. Pharyngobranchials with dental plates. The epi- and pharyngobranchials of the right side, in dorsal view. Photo L. Niven.

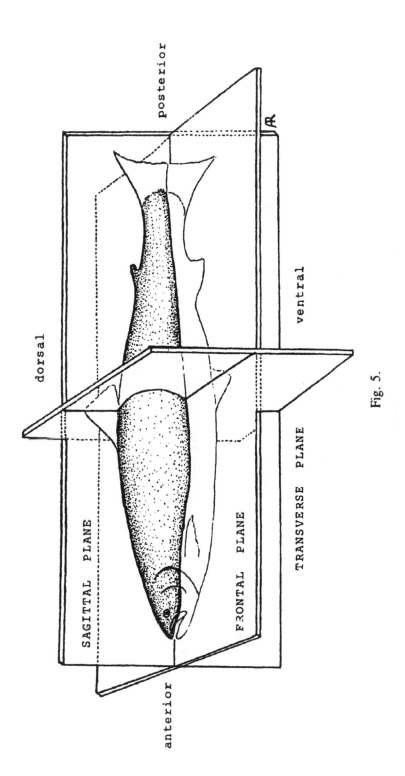

Fig. 5.

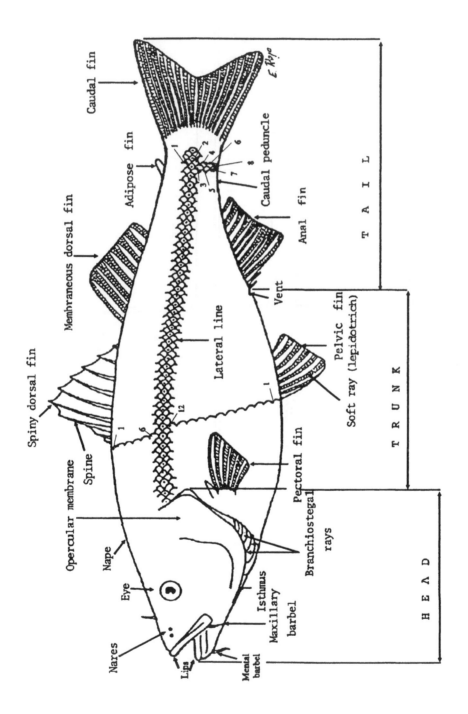

Fig. 6. External morphology of composite bony fish.

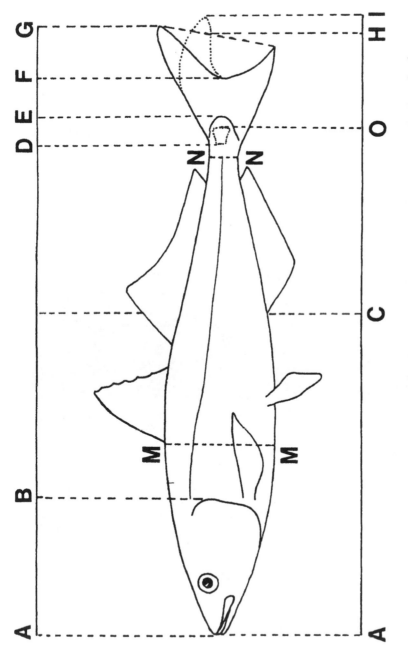

Fig. 7. Main morphometric characteristics of a bony fish: AB: Head length; AD: Standard length (to the beginning of the urostyle); AE: standard length (to the end of the caudal musculature); AF: Fork length; AG: Total normal length; A'C: Preanal length; A'H: Total auxillary length; A'I: Extreme tip length; MM': Maximum height; NN': Height of the caudal peduncle. (See **Total length and Standard length** in the Descriptive Section).

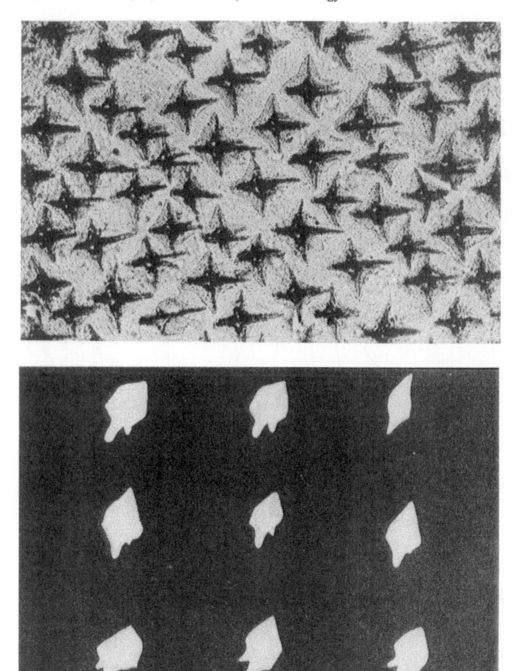

Fig. 8. **A**. Placoid scales of the dogfish (*Squalus acanthias*). **B**. Ganoid scales (*Lepisosteus*).
Photo A. Rojo.

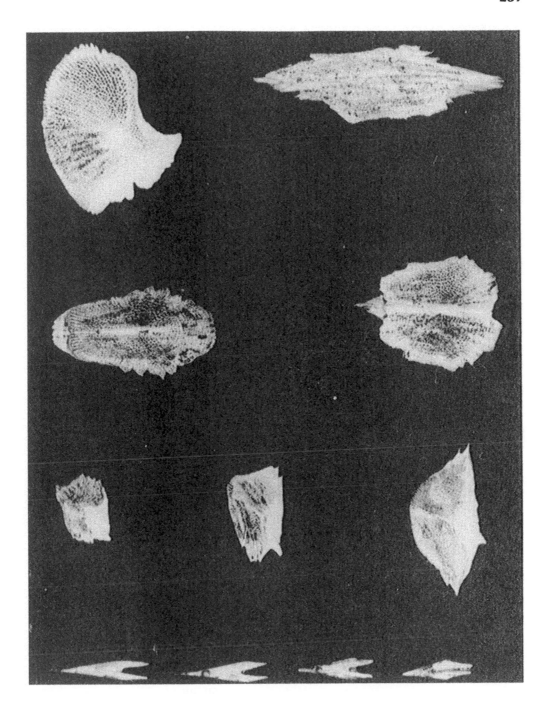

Fig. 9. Scutes of an Atlantic sturgeon (*Acipenser oxyrhynchus*) 1350 mm in length. Opercular (upper row left). Frontal (upper row right). Dorsal scutes (2nd row). Left lateral scuta (3rd row). Fulcra from the upper edge of the caudal fin (lower row). Photo L. Niven.

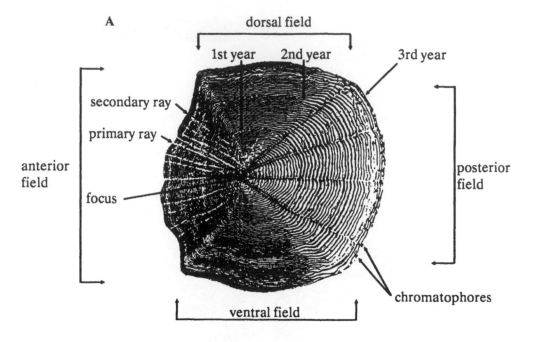

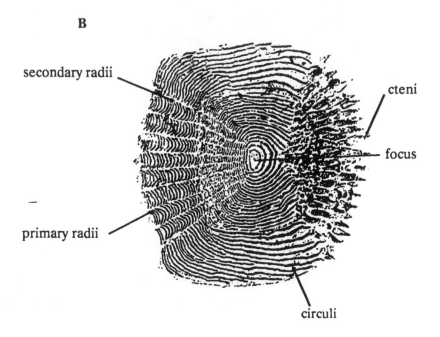

Fig. 10. Morphological features of a scale. **A.** cycloid scale. **B.** ctenoid scale

239

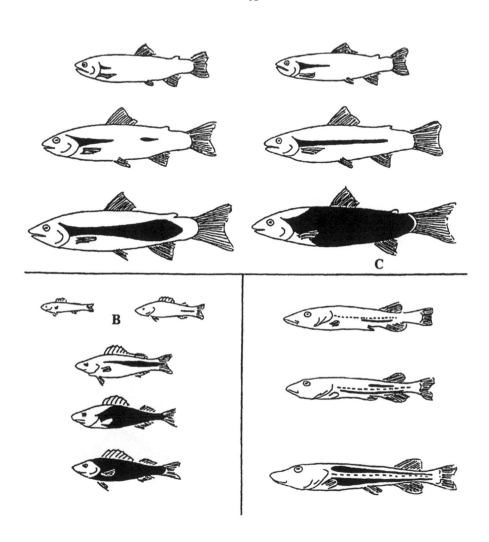

Fig. 11. Lepidotaxia. Three of the most common patterns of scale appearance and body covering. A Atlantic salmon (Rojo and Ramos, 1983). B. Walleye (Franklin and Smith, 1960). C. Pike (Priegel, 1964).

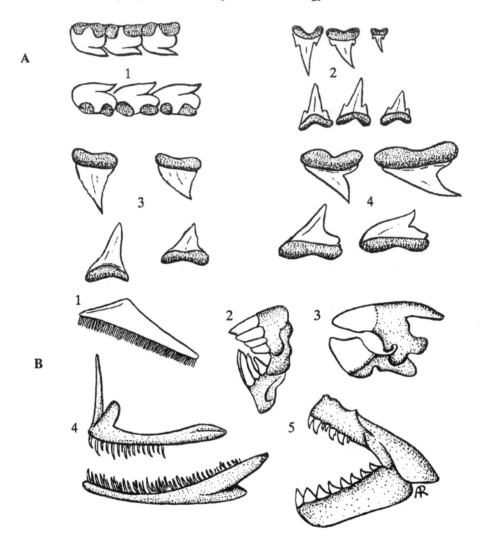

Fig. 12. Homodonty. A. Caniniform teeth in sharks. 1. Upper and lower teeth of the spiny dogfish. 2. Upper and lower teeth of the mackeral shark (*Lamna nasus*). 3. Blue shark teeth (*Prionace glauca*). 4. Hammerhead shark's teeth (*Sphyrna zygaeus*). Courtesy of the U.S. Department of the Interior. Fish and Wildlife Service. B. Types of bony fish teeth. 1. Cardiform teeth (*Prototrocles*). 2. Incisiform teeth (*Balistes caprus*). 3. Dental plate (*Scarus vetula*). 4. Caniniform teeth (*Lophius americanus*). 5. Piranha's teeth (*Serrasalmus*).

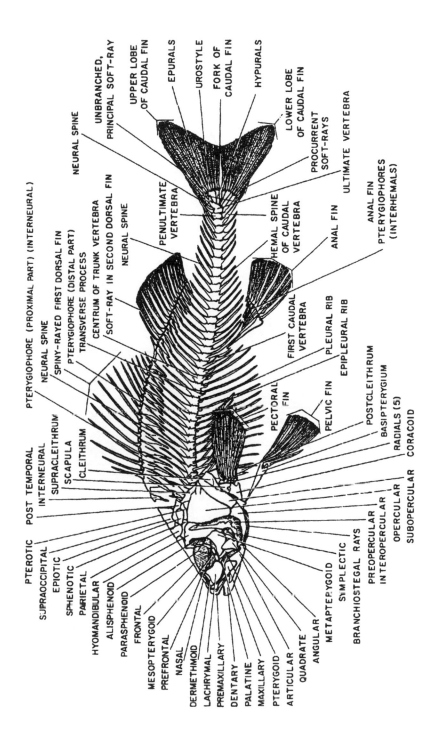

Fig. 13. Skeleton of a spiny-rayed bony fish (*Perca flavescens*). The name alisphenoid should be replaced by *pterosphenoid*. Courtesy of the late Dr. K. F. Lagler.

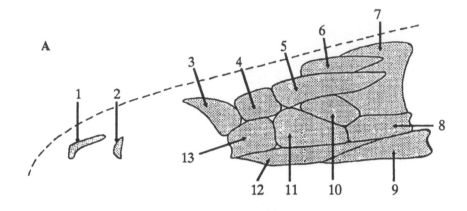

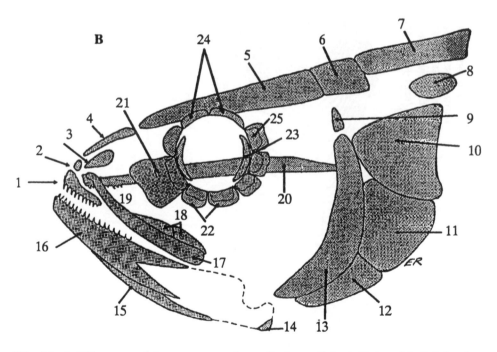

Fig. 14. A. Diagram of the neurocranium of a bony fish. 1. Ethmoid. 2. Myodome. 3. Orbitosphenoid. 4. Sphenotic. 5. Pterotic. 6. Epiotic. 7. Supraoccipital. 8. Exoccipital. 9. Basioccipital. 10. Opisthotic. 11. Prootic. 12. Basisphenoid. 13. Pterosphenoid. B. Diagram of the dermocranium of a bony fish. 1. Premaxilla. 2. Rostral. 3. Lateral ethmoid. 4. Nasal. 5. Frontal. 6. Parietal. 7.Supraoccipital crest. 8.Tabular. 9. Suprapreopercle. 10. Opercle. 11. Subopercle. 12. Interopercle. 13. Preopercle. 14. Retroarticular. 15. Gular plate. 16. Dentary. 17. Maxilla. 18. Supramaxillae. 19. Prevomer. 20. Parasphenoid. 21. Lacrymal. 22. Infraorbitals (2nd to 6th [23 = dermosphenotic]). 24. Supraorbitals. 25. Sclerotic

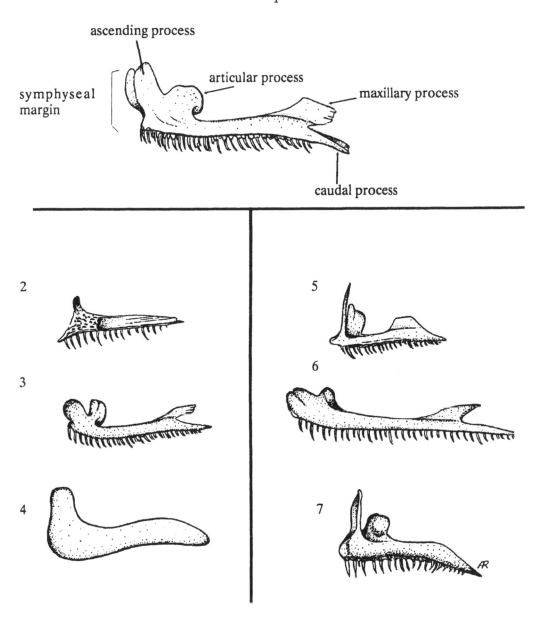

Fig. 15. Premaxilla. 1. Cod [82 mm]. 2. Salmon (*Salmo salar*) [34 mm]. 3. Cusk (*Brosme brosme*) [37 mm]. 4. Barb (*Barbus bocagei*) [15 mm]. 5. *Lutjanus* [28 mm]. 6. Argentinian hake (*Merluccius hubbsi*) [63 mm]. 7. Halibut (*Hippoglossus hippoglossus*) [40 mm].

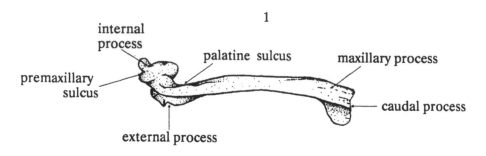

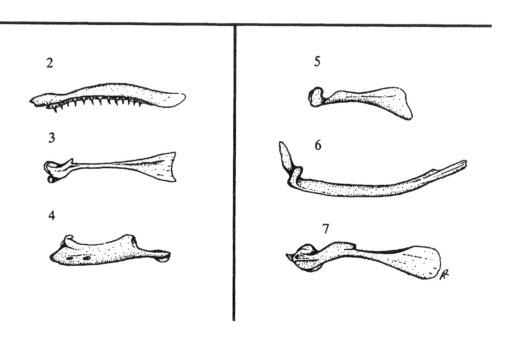

Fig. 16. Maxilla. 1. Cod [78mm]. 2. Salmon [62 mm]. 3. *Caranx hippos* [30 mm]. 4. Barb [22 mm]. 5. *Lutjanus* [32 mm]. 6. Angler (*Lophius americanus*) [114mm]. 7. Halibut [49mm].

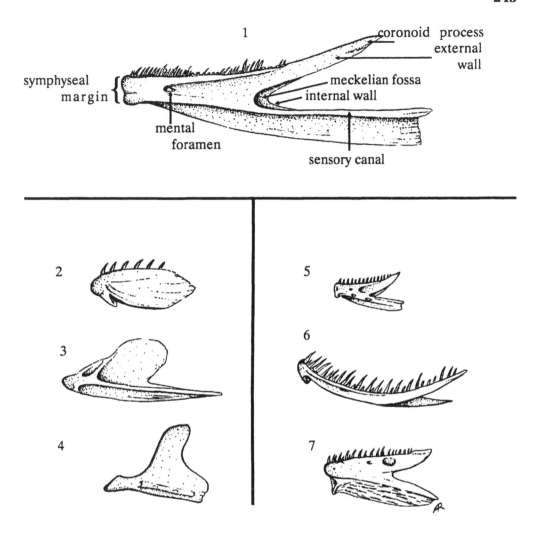

Fig. 17. Dentary. 1. Cod [90 mm]. 2. Salmon [28 mm]. 3. American shad [*Alosa sapidissima*) [27 mm]. 4. Barb [17 mm]. 5. *Lutjanus* [29 mm]. 6. Angler [102 mm]. 7. Halibut [49 mm].

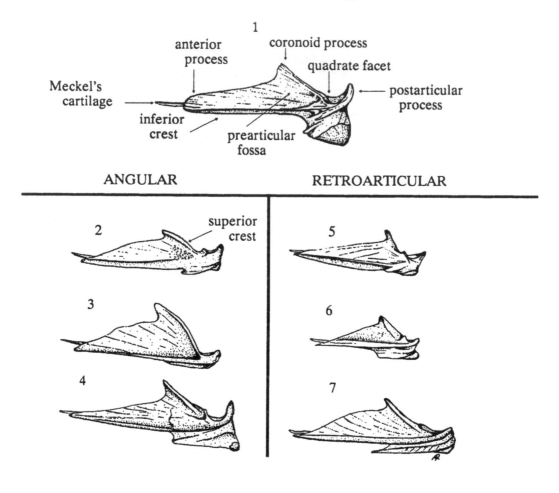

Fig. 18. Angular and Retroarticular. 1. Cod [77 mm]. 2. Salmon [42 mm]. 3. American shad [45 mm]. 4. Halibut [54 mm]. 5. Ocean pout (*Macrozoaroes americanus*) [43 mm]. 6. *Lutjanus* [34 mm]. 7. Argentinian hake [57 mm].

1

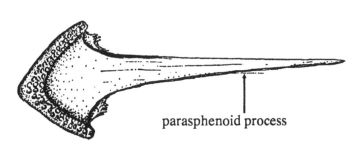

parasphenoid process

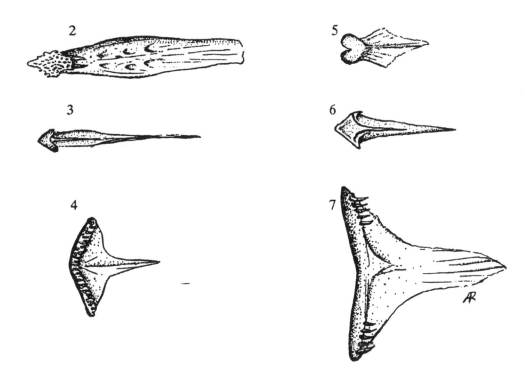

Fig. 19. Prevomer. 1. Cod [98 mm]. 2. Salmon [46 mm] 3. American shad [45 mm].
4. Toadfish (*Halobatrachus didactylus*) [28 mm]. 5. White sucker (*Catostomus commer-soni*) [20 mm]. 6. *Lutjanus* [30 mm]. 7. Angler [49 mm].

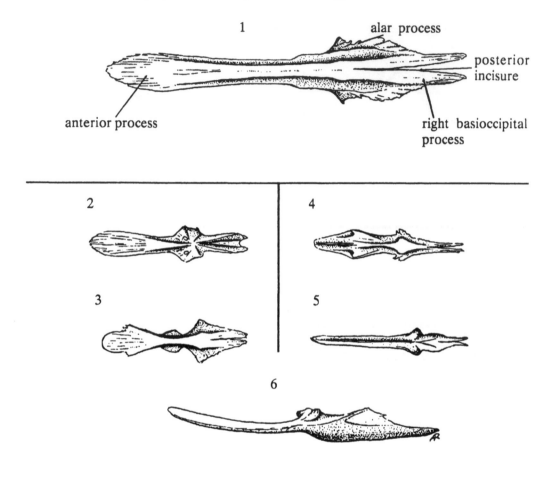

Fig. 20. Parasphenoid. 1. Cod [155 mm]. 2. Salmon [51 mm]. 3. Toadfish [46 mm].   4. White sucker [46 mm]. 5. *Lutjanus* [52 mm]. 6. American shad [75 mm].

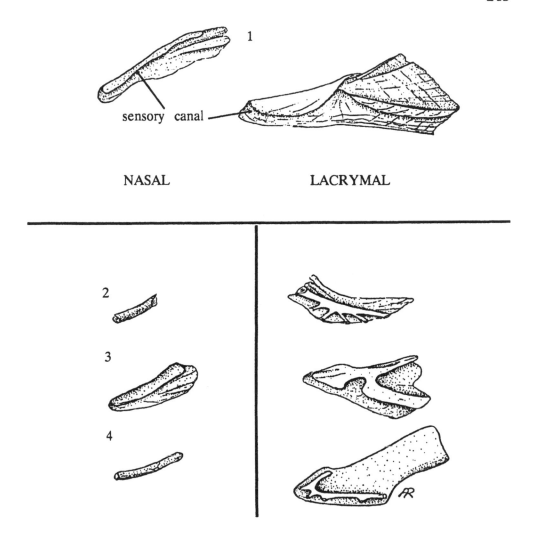

NASAL   LACRYMAL

sensory canal

Fig. 21. Nasal and Lacrymal. 1. Cod [46 and 71 mm]. 2. Arctic grayling (*Thymallus arcticus*) [15 and 35 mm] (after Norden, 1961). 3. Barb [12 and 40 mm]. 4. Haddock (*Melanogrammus aeglefinus*) [24 and 42 mm].

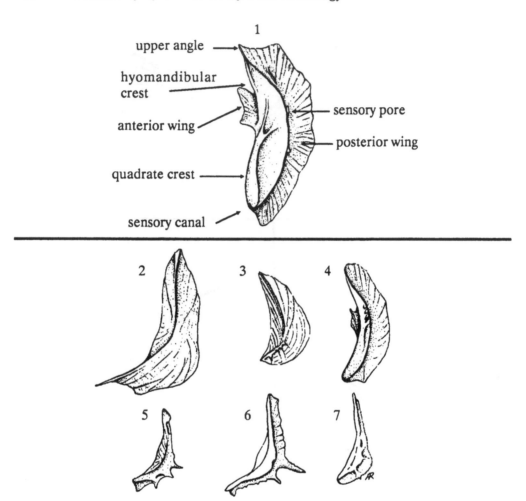

Fig. 22. Preopercle. 1. Cod [85 mm]. 2. Salmon [70 mm]. 3. American shad [53 mm]. 4. Haddock [58 mm]. 5. Redfish (*Sebastes marinus*) [23 mm]. 6. *Elagatis bipinnula*. 7. *Lutjanus* [52 mm].

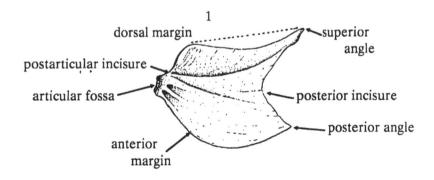

1

dorsal margin

superior
angle

postarticular incisure

articular fossa

posterior incisure

posterior angle

anterior
margin

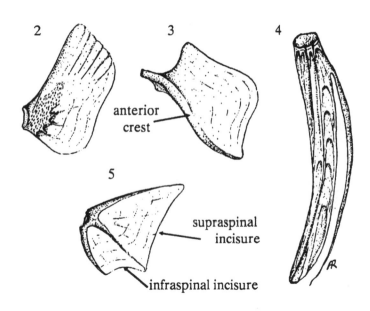

2

3

anterior
crest

4

5

supraspinal
incisure

infraspinal incisure

Fig. 23. Opercle. 1. Cod [42 mm]. 2. American shad [52 mm]. 3. Barb [35 mm]. 4. Angler [95 mm]. 5. *Lutjanus* [30 mm].

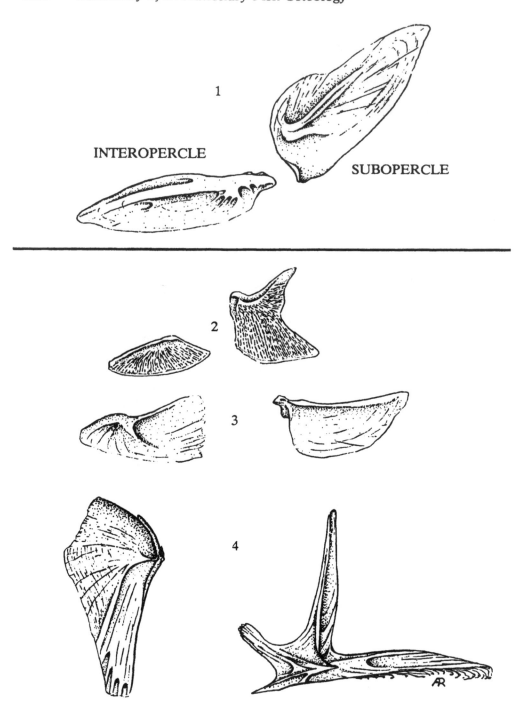

Fig. 24. Interopercle and Subopercle. 1. Cod [57 and 52 mm]. 2. Bowfin (*Amia calva*) [25 and 25 mm]. 3. Salmon [39 and 43 mm]. 4. Angler [60 and 76 mm].

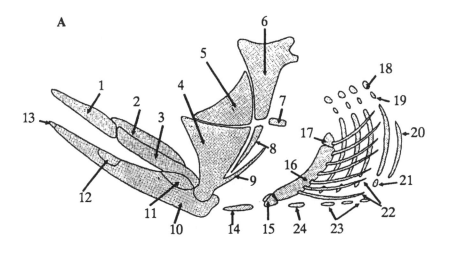

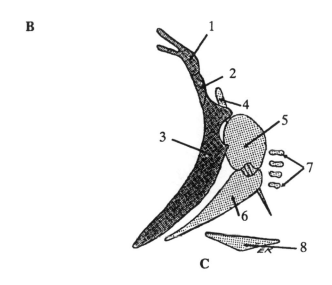

Fig. 25. Diagram of the splanchnocranium with some associated dermal bones. **A**. 1. Palatine. 2. Endopterygoid. 3. Ectopterygoid. 4. Quadrate. 5. Matapterygoid. 6. Hyomandibular. 7. Interhyal. 8. Symplectic. 9. Quadratojugal process. 10. Angular. 11. Coronomeckelian. 12. Mediomeckelian. 13. Mentomeckelian. 14. Basihyal (= Glossohyal). 15. Hypobrachials (dorso- and ventrohyal). 16. Ceratohyal (= anterohyal). 17. Epihyal (= posterohyal). 18. Pharyngobranchials. 19. Epibranchials. 20. Ceratobranchials. 21. Hypobranchials. 22. Branchiostegal rays. 23. Basibranchials. 24. Urohyal. **B**. Diagram of the secondary pectoral girdle (dark stippled) and primary pectoral girdle (light stippled). 1. Posttemporal. 2. Supracleithrum. 3. Ceithrum. 4. Postcleithrum. 5. Scapula. 6. Coracoid. 7. Radials. **C**. Pelvic girdle. 8. Basipterygium.

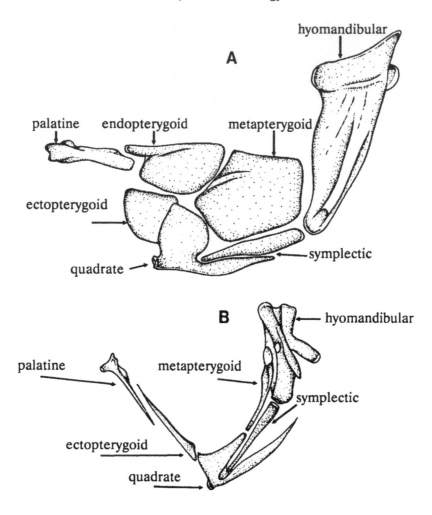

Fig. 26. Suspensorium. A. Barb (Rojo, 1987). B. *Aphya minuta* (Rojo, 1985).

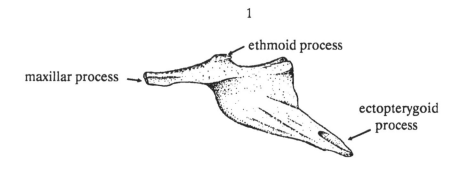

1

maxillar process →

ethmoid process

ectopterygoid
process

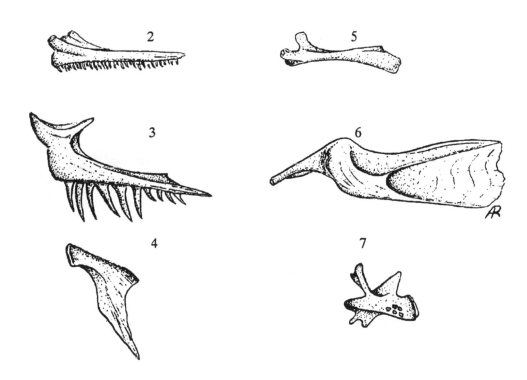

2

5

3

6

4

7

Fig. 27. Palatine. 1. Cod [62 mm]. 2. Toadfish [22 mm]. 3. Angler [57 mm]. 4. Halibut [32 mm]. Barb [14 mm]. 6. Argentinian hake [36 mm]. 7. White sucker [14 mm].

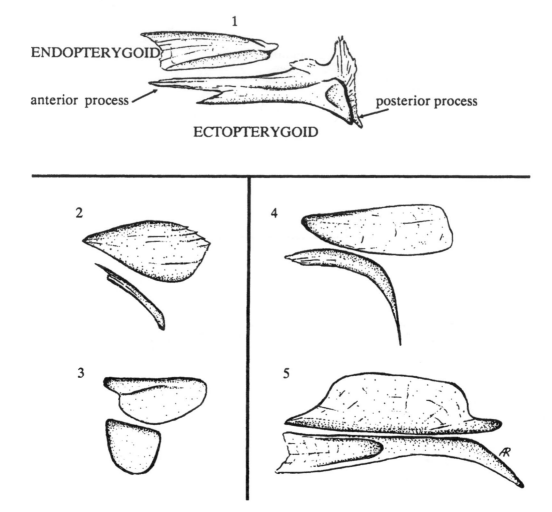

Fig. 28. Endopterygoid and Ectopterygoid. 1. Cod [43 and 58 mm]. 2. Arctic grayling (after Norden, 1961). 3. Barb [11 and 12 mm]. 4. *Lutjanus* [25 and 21 mm]. 6. Argentinian hake (*Merluccius hubbsi*) [37 and 42 mm].

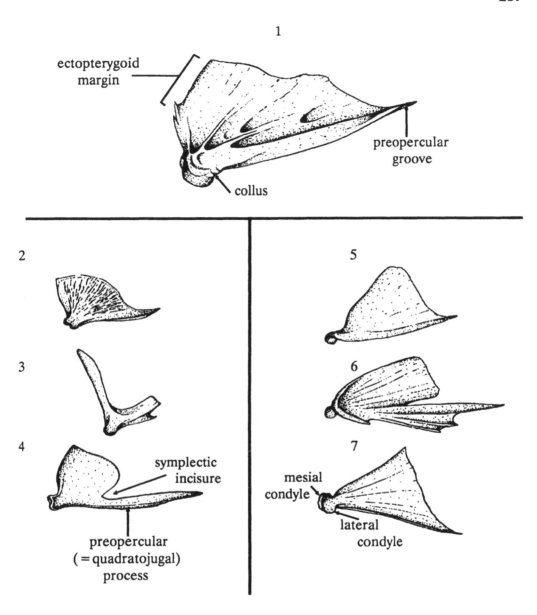

Fig. 29. Quadrate. 1. Cod [50 mm]. 2. Salmon [15 mm]. 3. American shad [15 mm]. Barb [15 mm]. 5. *Lutjanus* [20 mm]. 6. Angler [60 mm]. 7. Halibut [57 mm].

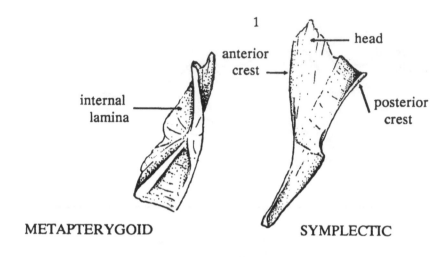

1

internal
lamina

anterior
crest

head

posterior
crest

METAPTERYGOID   SYMPLECTIC

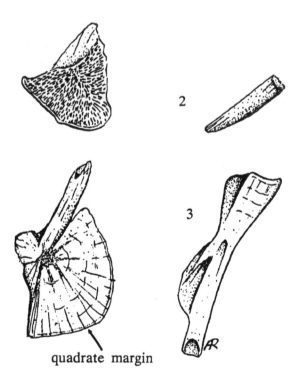

2

3

quadrate margin

Fig. 30. Metapterygoid and Symplectic. 1. Cod [35 and 49 mm]. 2. Salmon [27 mm]. 3. Angler [48 and 52 mm].

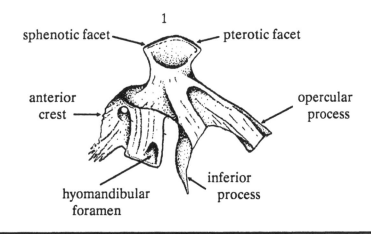

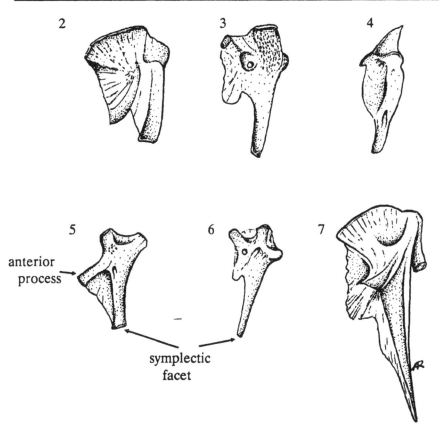

Fig. 31. Hyomandibular. 1. Cod [48 mm]. 2. Salmon [27 mm]. 3. Toadfish [25 mm]. 4. American shad [43 mm]. 5. *Lutjanus* [28 mm]. 6. Barb [30 mm]. 7. Angler [67 mm].

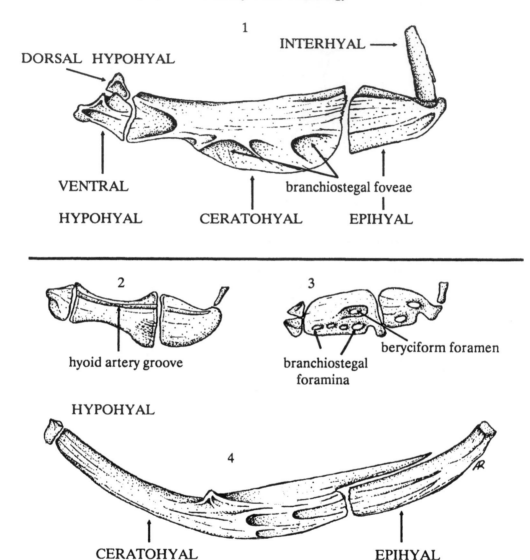

Fig. 32. Hyoid arch. 1. Cod [110 mm]. 2. Salmon [45 mm]. 3. *Pomolobus mediocris* (after Caillet, *et al* 1986). 4. Angler [150 mm].

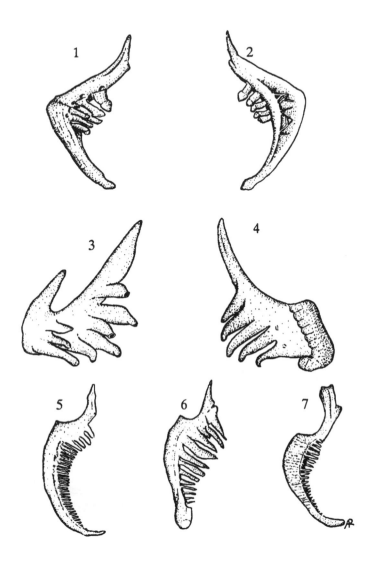

Fig. 33 . Lower pharyngeal (= 5th ceratohyal). 1. Left, dorsal view. Barb [27 mm].
2. Ventral view of same. 3. Left, dorsal view. Creek chub (*Semotilus atromaculatus*) [17 mm]. 4. Left, ventral view. Creek chub [10 mm] (after Evans and Deubler, 1955).
5. Right, ventral view of juvenile largescale sucker (*Catostomus macrocheilus*) [2.5 mm]. 6. Right, ventral view of an adult largescale sucker [32 mm] (after Weisel, 1967).
7. Left, dorsal view. White sucker [30 mm].

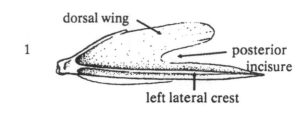

GLOSSOHYAL   UROHYAL

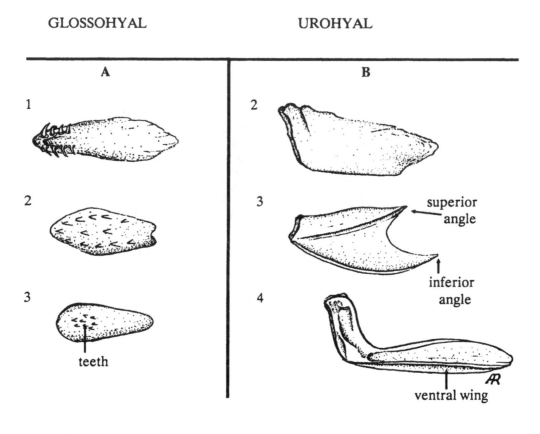

Fig. 34. **A** . Glossohyal. 1. Salmo [29 mm]. 2. *Coregonus clupeaformis*. 3. Arctic grayling (after Norden, 1961). **B**. Urohyal. 1. White sucker [32 mm]. 2. Cod [33 mm]. 3. *Lutjanus* [30 mm].   Argentinian hake [47 mm]

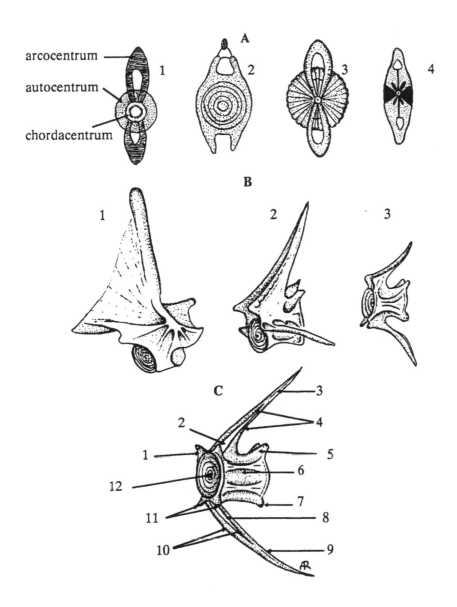

Fig. 35. Monospondylous vertebrae. A. Shark vertebrae. 1. Embryonic elements. 2. Cyclospondylous vertebra. 3. Asterospondylous vertebra. B. Amphicelous fish vertebrae of cod. 1. Atlas. 2. Precaudal vertebra. 3. Caudal vertebra. C. Morphology of a typical caudal vertebra. 1). Anterior dorsal prezygapophyses. 2). Neural canal. 3). Neural spine. 4). Neural arch. 5). Posterior dorsal postzygapophyses. 6). Centrum. 7). Posterior ventral postzygapophyses. 8). Hemal canal. 9). Hemal spine. 10). Hemal arch. 11). Anterior ventral prezygapophyses. 12). Canal for the passage of the notochordal tissue.

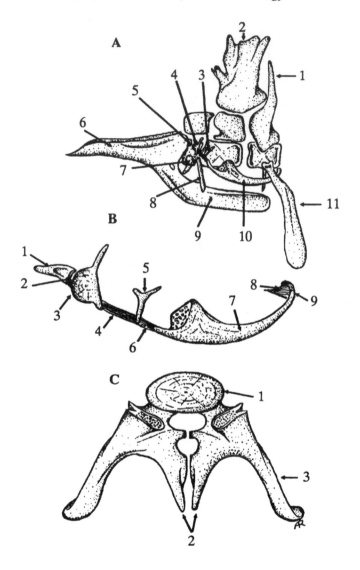

Fig. 36. **A.** Neural complex of barb [35 mm]. 1. Neural arch of the 4th vertebra. 2. Neural complex. 3. Intercalarium. 4. Scaphium. 5. Claustrum. 6. Basioccipital. 7. Lateral process of the 1st vertebra. 8. Lateral process of the 2nd and 3rd vertebrae. 9. Pharyngeal process of the basioccipital. 10. Tripus. 11. Process of the 4th vertebra. **B.** Left series of the Weberian ossicles [24 mm]. 1. Claustrum. 2. Claustro-scaphial ligament. 3. Scaphium. 4. Scaphio-intercalar ligamentum. 5. Intercalarium. 6. Intercalo-tripodal ligamentum. 7. Tripus. 8. *Tensor tripodis.* 9. Transformator process. **C.** Posterior view of the suspensorium [20 mm]. 1. Centrum of the 4th vertebra. 2. *Ossa suspensoria.* 3. Process of the 4th vertebra.

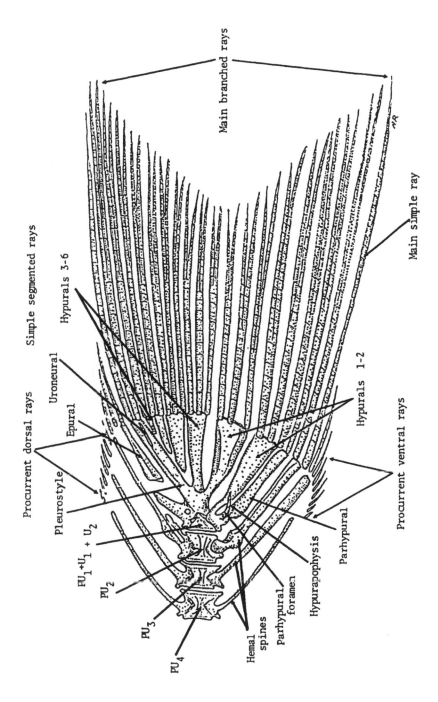

Fig. 37. Caudal skeleton and caudal fin of barb [105 mm]. See Caudal Skeleton in the Descriptive Section.

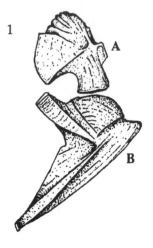

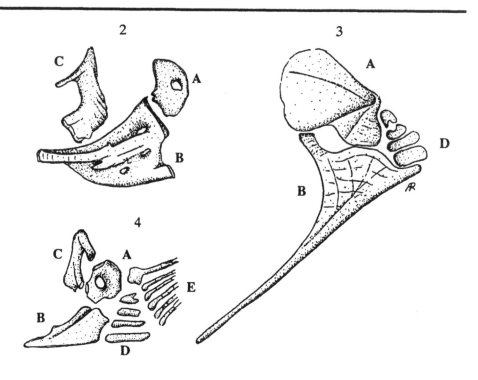

Fig. 38. Pectoral girdle. A. Scapula. B. Coracoid. C. Mesocoracoid. D. Radials. E. Fin rays (lepidotrichia). 1. Cod. 2. Arctic grayling (after Norden, 1961). 3. Argentinian hake [50 mm]. 4. Barb (Rojo, 1987).

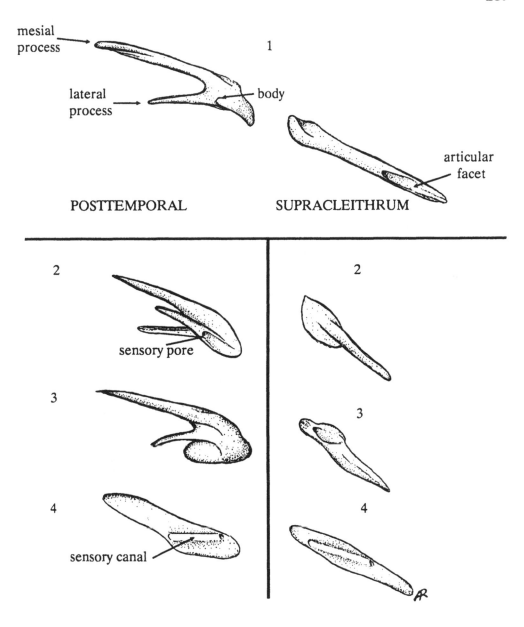

Fig. 39. A. Posttemporal. 1. Cod [50 mm]. 2. American shad [40 mm]. 3. Haddock [31 mm]. 4. Barb [25 mm]. B. Supracleithrum. 5. Cod [55 mm]. 6. American shad [35 mm]. 7. Haddock [31 mm]. 8. Barb [22 mm].

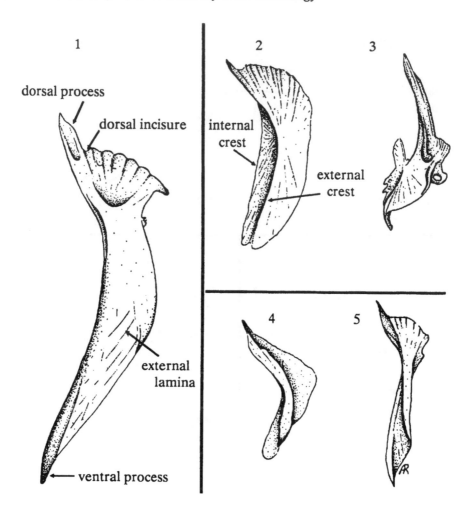

Fig. 40. Cleithrum. 1. Cod [149 mm]. 2. Salmon [56 mm]. 3. American shad [55 mm]. 4. Barb [45 mm]. 5. Lutjanus [59 mm].

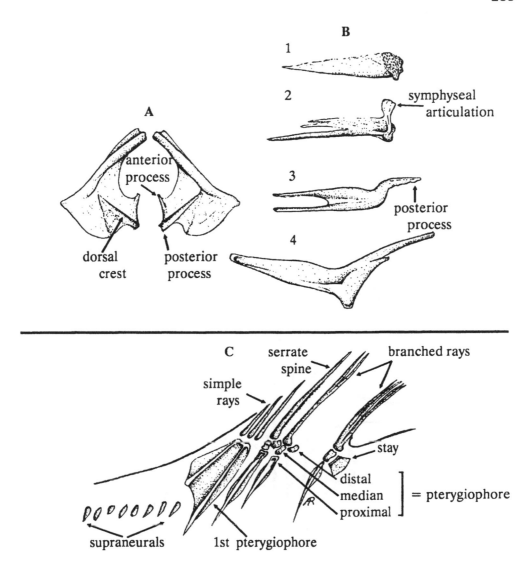

Fig. 41. **A**. Basipterygia. 1. Cod [32 mm]. 2. American shad [44 mm]. 3. White sucker [46 mm]. 4. Barb [23 mm]. 5. Argentinian hake [47 mm]. **B**. Pterygiophores of the dorsal fin of barb. All in dorsal view.

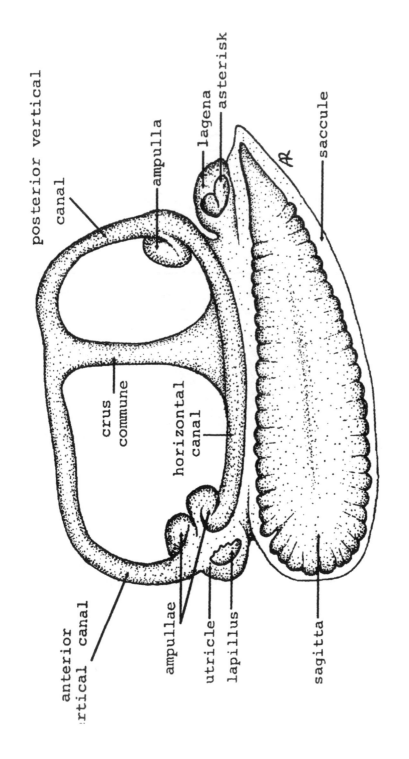

Fig. 42. A. The membranous labyrinth of the Argentinian hake.

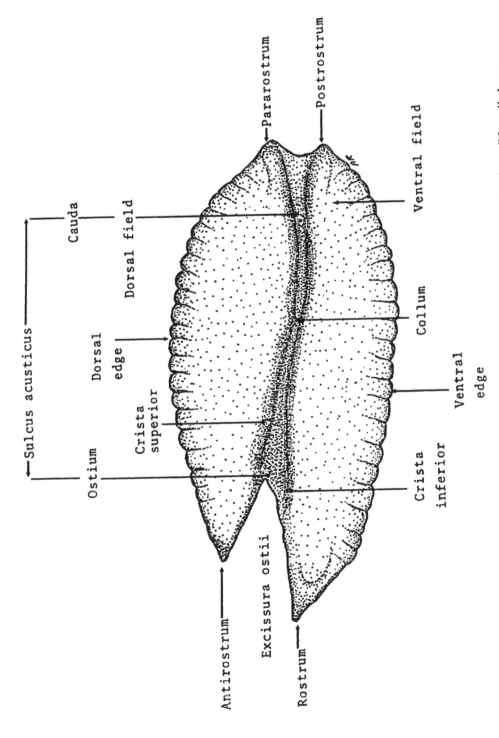

Fig. 43. A. Morphological features of a composite sagitta. Inner right view. (Not all these morphological features are present in all sagittae).

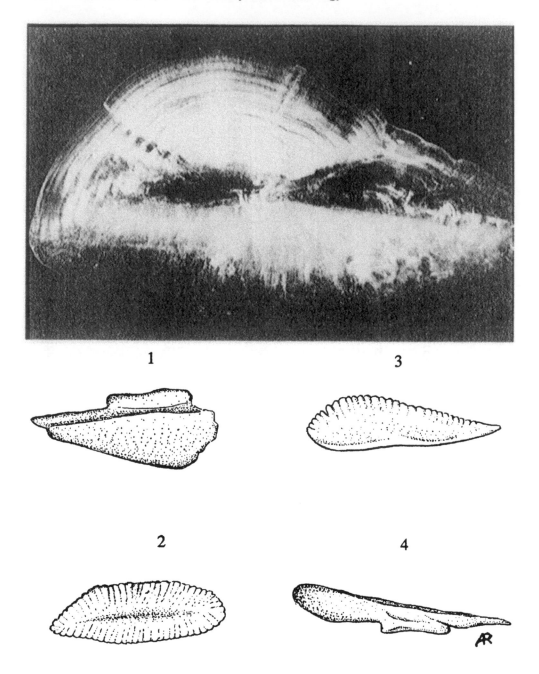

Fig. 44. **Upper section**. Annual rings of lake trout (*Cristivomer namaycush*). Photo A. Rojo. **Lower section**. Diversity of shapes of sagittae. 1. Inner face of the right sagitta of salmon (after Casteel, 1976). 2. Outer face of the left sagitta of cod [16 mm]. 3. Outer face of the left sagitta of the Argentinian hake [24 mm]. 4. Ventrolateral view of the sagitta of *Brycon* (after Weitzman, 1962).

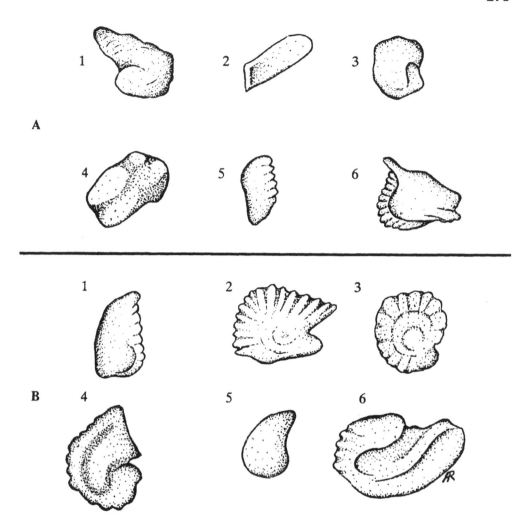

Fig. 45. *A*. Left view of lapilli. 1. Cod [2mm]. 2. Outer view, barb [2mm]. 3. Inner view, *Gobio gobio* (after Vanderwalle, 1975). 4. Outer view, *Brycon* (after Weitzman, 1962). 5. Outer view, Argentinian hake [2.5mm]. 6. *Scorpaena dactyloptera* (after Gottbehut, 1935). **B**. Asterisks. 1. Dorsal view of the left asterisk of cod [1.5mm]. 2. Left asterisk of barb [3.5mm]. 3. Inner face of right asterisk of *Gobio gobio* (after Vandewalle, 1975). 4. Inner face of right asterisk of *Brycon* (after Weitzman, 1962). 5. Left asterisk of the Argentinian hake [2.5mm]. 6. Inner face of right asterisk of *Peristedion* (after Gottbehut, 1935).